Kister Distillation Symposium 2023

Topical Conference at the 2023 AIChE Spring Meeting and 19th Global Congress on Process Safety

Houston, Texas, USA
12-16 March 2023

ISBN: 978-1-7138-7937-4

Printed from e-media with permission by:

Curran Associates, Inc.
57 Morehouse Lane
Red Hook, NY 12571

Some format issues inherent in the e-media version may also appear in this print version.

Printed with permission by Curran Associates, Inc. (2023)

For permission requests, please contact AIChE
at the address below.

AIChE
120 Wall Street, FL 23
New York, NY 10005-4020

Phone: (800) 242-4363
Fax: (203) 775-5177

www.aiche.org

Additional copies of this publication are available from:

Curran Associates, Inc.
57 Morehouse Lane
Red Hook, NY 12571 USA
Phone: 845-758-0400
Fax: 845-758-2633
Email: curran@proceedings.com
Web: www.proceedings.com

TABLE OF CONTENTS

KISTER DISTILLATION SYMPOSIUM 2023: LEGACY OF JOSE BRAVO: BREAKING TECHNOLOGIES FOR OVERCOMING THE ENERGY TRANSITION CHALLENGES IN DISTILLATION

KISTER DISTILLATION SYMPOSIUM 2023: DIVIDING WALL COLUMNS: THE RAPIDLY EXPANDING TECHNOLOGY AND OTHER EXCITING TECHNOLOGIES LEADING THE WAY TO ENERGY TRANSITION

KISTER DISTILLATION SYMPOSIUM 2023: HONORING THE CONTRIBUTIONS OF ELIZABETH AND NORMAN LIEBERMAN: TROUBLESHOOTING, PRACTICAL ENGINEERING, AND INSPIRATION TO ENGINEERS

KISTER DISTILLATION SYMPOSIUM 2023: LEGACY OF DR. WALTER STUPIN: DISTILLATION SIMULATIONS ARE ONLY GOOD WHEN THE VLE AND PHYSICAL PROPERTIES PREDICTIONS ARE GOOD

KISTER DISTILLATION SYMPOSIUM 2023: NEW INSIGHTS INTO DISTILLATION CONTROL, REBOILERS, AND CONDENSERS

KISTER DISTILLATION SYMPOSIUM 2023: TROUBLESHOOTING AND REVAMP IN ETHYLENE PLANT, GAS PROCESSING, PETROCHEMICAL, AND REFINERY COLUMNS

KISTER DISTILLATION SYMPOSIUM 2023: TROUBLESHOOTING: THE LESSONS LEARNED APPLIED TO BETTER OPERATION

94b The Potential Impact of the Energy Transition on Process Plants and in Particular Distillation Units

Izak Nieuwoudt, Tony Cai

The potential impact of the energy transition on process plants and in particular distillation units

Izak Nieuwoudt & Tony Cai

Fractionation Research Inc.

The fear of climate change and its impact, and the probable correlation with anthropogenic greenhouse emissions are driving an energy source transition that is affecting the chemical process industries. With distillation being the largest energy consumer in these industries, these factors are starting to have an impact on how distillation is perceived. Technologies that can reduce the energy consumption of separation processes can make a significant contribution to the decarbonization of distillation units. Driving these separation processes with low-carbon energy sources will further address the concerns. In this presentation the authors will discuss high efficiency distillation equipment, reduced energy consumption distillation systems, the use of alternative energy sources and the potential overall impact on the decarbonization of separation systems employing distillation.

Keywords: Distillation, Energy, Decarbonization, Efficiency

94c Design and Characterization of a Flexible Tray Column with Enhanced Operating Window and Turndown Ratio

Julia Riese, Niklas Breuer, Marcus Gruenewald

Ensuring future competitiveness of production processes is an important task for the chemical industry. This requires the processes to be constantly and adequately adapted for the related challenges. These challenges result, among others, from the need to reduce greenhouse gases significantly. In particular, future energy and feedstock supply will increasingly rely on renewable rather than on fossil resources. Especially, an energy supply structure that heavily relies on intermittent renewable resources imposes new challenges for the operation of chemical production processes.

Adjusting the operation of the processes to the availability of energy streams is possible by enhancing the flexibility of processes and their unit operations In literature, various solution strategies to enhance process flexibility can be found, e.g. for chemical reactors. This also applies to downstream processing unit operations such as distillation columns. In fact, distillation processes consume a considerable amount of energy, so that energy- and separation-optimal operation needs to be ensured at all operating points. Technical solutions are required to cope with these increasing uncertainties.

In our work, we have developed a novel column design that allows for a more flexible operation of distillation processes. The internal volume of the column is segmented by installing vertical dividing walls throughout the total height of the column. Every segment in the column can be operated, started-up and shut-down independently. In fact, each segment can be operated at a distinctive separation-optimal operating point, which is the basis for the increased flexibility.

The segmentation of the column resulted in two specific tray layouts to achieve efficient contact between gas and liquid on each tray. The tray layouts differ in the flow direction of the liquid over the active area of the tray and therefore in the positioning of the weir and the downcomer. In the tray layout D_I, the liquid flows from an inner laying downcomer in radial direction over the active area to an outer weir. On the tray with the layout D_O, the flow direction of the liquid is vice versa and the liquid is entering the active area from an outer laying downcomer towards an inner weir. Both tray layouts are based on sieve trays for the active area.

Besides these distinctive tray layouts, the column is characterized by the capability of coupling the segments on each tray via the downcomer next to each other. The motivation behind this coupling is to accelerate start-up operation of single segments. If one segment is operated at steady-state, the start-up of the adjacent segment can be supported by liquid transferred from the downcomer of each tray. Thus, an inactive segment is started-up using a pre-defined concentration profile. In addition, we developed a dynamic column model to perform start-up simulations, analyzing this approach to start-up individual segments within one column shell. Results show that a significant reduction of the start-up time is possible, while the active segment is affected only for a short time. During start-up, only <10% of the liquid holdup from the active downcomer was used. Anyway, to validate these simulation results, it is necessary, to assess the liquid transfer from one active downcomer to the adjacent inactive ones in experimental studies. In addition, insights into the specific operation of single segments in this column design are yet to be discovered.

In this contribution, we present the results of the experimental characterization of the innovative tray design focusing on (a) characterization of fluid dynamics and operating limits and (b) strategies for a flexible operation of the new tray design in a d_{column}=600 mm test column. The column consists of four segments and can be operated with gas loads up to 2 $Pa^{0.5}$ and liquid flow rates up to 5.5 m^3 per hour. Usually, for sieve trays, the liquid load is defined per meter weir length. For the particular tray design used in the flexible column, however, liquid load is specified per m^2 active area to ensure comparability between the tray layouts. This results in liquid loads between 25 $m^3 \cdot m^{-2} \cdot h^{-1}$ and 105 $m^3 \cdot m^{-2} \cdot h^{-1}$.

For the characterization of fluid dynamics, we determined dry and total pressure drops for different tray design parameters. Here, design parameters vary in terms of hole diameter and resulting opening ratio. We compare experimental results with existing correlations from literature. In addition, we assessed the weep point to define the lower operating boundary. Due to the different described tray layouts (D_I and D_O), flow regimes and liquid distribution on the trays is of special interest. Therefore, we present results from ultrasonic measurements of liquid heights on the trays. The ultrasonic sensors were distributed over the active area of the tray layouts to assess the liquid height at different locations.

Secondly, we operated the column under liquid transfer from one segment to an adjacent one to enable flexible operation. Therefore, we assessed maximum transfer rates. Gaining insights into operating stability was possible by pressure drop measurements on the trays and in the downcomers. Preforming the measurements at varying gas and liquid loads leads to the identification of superimposed operating boundaries during the operation with liquid transfer. In addition, the results help to determine switching points for the flexible operation. Switching points define the start-up or shut down of individual segments in the course of flexible operation.

Lastly, we operated two adjacent segments in parallel and assessed fluid dynamics experimentally. By operating two adjacent segments, we show that out design is robust and parallel operation of segments is based on comparable conditions.

94d Unraveling the Multifaceted Challenges and Advances in Validating Tray Efficiency Prediction Models

Vineet Vishwakarma, Markus Schubert, Uwe Hampel

Industrial tray columns are widely used for distillation and absorption processes globally. They are known for high energy consumption, which is often overlooked due to unavailability of an equivalent industrially viable alternative. Rising energy costs and urgent need to control greenhouse gas emissions call for improvement in the performances of tray columns globally. This can be achieved by tuning the dynamics of the two-phase dispersion on individual trays for higher efficiencies via design modifications and revamping. To do so, it becomes necessary to understand how the two-phase flow evolves over a tray and relates to the tray efficiency. Such relation can be evaluated based on mathematical models called as tray efficiency prediction models. Hitherto, the existing models only provided black box estimations and ignored maldistributions in the vapor flow. These limitations were recently targeted by a new model referred to as 'Refined Residence Time Distribution (RRTD) model'.[1]

The proof of concept of the RRTD model demands complete information pertaining to two-phase dispersion and mass transfer on a large-scale column tray. They are not available at desired resolution in the existing literature due to several limitations in the applied measurement techniques and systems. Thus, a recently-proposed multiplex flow profiler[2] was deployed inside an air-water column mockup (DN800) for characterizing the distributions of liquid holdup, residence time and mixing over a sieve tray for several loadings at high resolution. For the same operating conditions, the efficiency data over that tray was obtained based on air-led stripping of isobutyl acetate from the aqueous solution. Both hydrodynamic and efficiency data were applied together for assessing the validity of the new RRTD and other models. This works also sets new benchmarking standards for improved validation of CFD and efficiency prediction models in the future.

[1] Vishwakarma, V., Schubert, M. and Hampel, U., 2019. Development of a refined RTD-based efficiency prediction model for cross-flow trays. *Industrial & Engineering Chemistry Research*, 58(8), pp.3258-3268.

[2] Vishwakarma, V., Schleicher, E., Schubert, M., Tschofen, M. and Löschau, M., 2020. Sensor zur Vermessung von Strömungsprofilen in großen Kolonnen und Apparaten. *Deutsches Patent und Markenamt*, DE 10 2018 124 501.

94e CFD Analysis of Distillation Trays and Separator Drums: How Revealing a CFD Analysis Can be?

Babak Rafi

Use of Computational Fluid Dynamics (CFD) has become more common to analyze, validate a design, predict, and solve problems in applications that involves separation technologies and equipment such as trays, packings, mist eliminators, etc. CFD analysis provides great insight to how vapor and liquid phases behave in an operational unit giving visual insights. However, CFD has its own limitations. This paper will review a few troubleshooting case studies from industrial applications with unique and challenging 1-phase (vapor) and 2-phases (vapor/liquid) CFD analysis of distillation trays and separator drums. Additionally, it will discuss the importance of proper interpretation of the CFD results and required expertise in separation field to validate and see how revealing or misleading the results can be.

Experimental Investigations on the Fluid-Dynamic Design of High-Performance Separating Trays Based on a Miniaturized Measuring

Wibke Leushacke, Ruhr University Bochum, Bochum, Germany
Amelie Merkel, Ruhr University Bochum, Bochum, Germany
Marcus Gruenewald, Ruhr University Bochum, Bochum, Germany

Abstract

A design approach for a miniaturized measuring cell to determine the flow behavior of sieve and valve groups on trays is presented. The system is intended to improve the design process of tray columns. Measurements regarding pressure drop and weeping are performed in a rectangular facility (0.6x0.2m). The new test system will be evaluated for its suitability as a support for pilot plant columns to collect experimental data for design. Special attention will be given to optimize the system for the investigation of organic substitute systems. Hydraulic measurements for liquid loads between $u_L = 0 - 40\ m^3m^{-2}h^{-1}$ and gas loading between $F_{v,ac} = 0.2 - 2.5\ Pa^{0,5}$ for a relative hole area $\varphi = 0.106$ were presented for the water/air system. Furthermore, an analysis of the tray crossflow between an ID600-column and the rectangular shape is made. Therefore, an IR-camera is used to measure the surface temperature. The results show that the measuring cell can reproduce the behavior at ideal tray crossflow. In order to use it for the design, further considerations are necessary - e.g. to specifically influence the tray crossflow - to give advice for designing trays. The investigation of different tray types and material systems is also pending.

Introduction

One of the greatest tasks of our time is the conversion of the energy system to renewable energies. Along with this, the demands on process design are becoming even greater. Plants must be able to react to fluctuating amounts of available energy. This is a great challenge, especially with regard to thermal separation processes. Operating limits should be known very precisely in order to always be able to run the plants at the optimal point depending on the amount of energy. Due to the numerous different tray types, this means a considerable effort, especially for tray columns. While sufficient experimental data are available for sieve trays and some types of float and fixed valves, new and optimized trays have been established in recent years for which the behavior at the operating limits has not been sufficiently researched. The influence of different material systems also continues to be a research gap.

For this reason, a concept is being developed at the Ruhr-Universität Bochum that enables the collection of experimental data in the smallest possible unit. In a rectangular system, the hydrodynamic phenomena are mapped on a tray. The focus is on the investigation of substitute systems as well as on the provision of a facility that provides the basis for further optimization of the design of tray columns. In order to put the system into operation, measurements are first made with the substance system water/air for sieve trays. Comparisons with relevant correlations and evaluated literature will be used to evaluate the use of the system. Subsequently, the system is converted to high-performance trays in order to prepare it for use with organic substitute systems. This paper shows the results of the start-up and initial investigations with high performance trays. The focus is also on tray crossflow.

Theoretical Background

In order to design a tray column, a large number of hydraulic parameters need to be studied [1]. The main focus is on the determination of the fluid dynamic limits for the operating window. In order to be able to show these limits, first of all experimental data are necessary. Based on these data, correlations are developed that predict the operating limits. The influence of different substance systems is also initially only included via empirical equations. The design methods developed in this way are adopted in software to perform scale-ups.

Many of the calculations are based on investigations of sieve trays and the water/air material system. As mentioned before, new trays have emerged over time, which are considered well researched to date, but are subject to high uncertainties, especially at the load limits. In order to eliminate these uncertainties, experimental investigations are again necessary. The major disadvantage of using technical plants for these measurements are the highly costs associated with production costs for the trays, plant peripherals, supply chains and material requirements. In most cases, also the measurement of organic substitute systems is complicated due to safety precautions. One concept to overcome these disadvantages is the miniaturization of equipment. This involves recording experimental data in smaller test setups that map the relevant zones. For columns with structured packing, it has already been shown that this concept produces reproducible and verified measurements [2, 3, 4]. The miniaturization approach is being pursued at Ruhr University Bochum for the study of high-performance trays.

This paper will mainly focus on the start and first measurements of the new test set-up. Of the numerous hydraulic parameters mentioned above, the following are in main focus: Pressure drop and Weeping. To test the performance of the new system, the first measurements will be carried out with water/air-system and sieve trays. For validation, the results are compared with correlations from the literature. Table 1 shows the correlation which will be considered.

Table 1: *Measurements and Correlations to design tray columns*

Reference		Parameter	Column/tray	Material system
Badínský	[5]	p_{dry}	d_i = 0.4 to 1.0 m/kittle and ripple plates	various
Červenka	[6]	p_{dry}	d_i = 0.288 m/sieve data from literature	air, methane, argon, carbon oxide, freon 12
Leibson	[7]	p_{dry}	data from literature	various
Smith	[8]	p_{dry}	d_i = 0,07 m/sieve	air, carbon dioxide, methane, argon
Stichlmair	[9]	p_{dry} $h_{clear} \rightarrow p_{total}$	various	various
Benett	[10]	$h_{clear} \rightarrow p_{total}$	A = 0.152x0.61m²/sieve	air/water
Colwell	[11]	$h_{clear} \rightarrow p_{total}$	data from literature	various
Zuiderweg	[12]	$h_{clear} \rightarrow p_{total}$	FRI Data	
Lockett	[13]	Weeping	A = 1.22x0.63m²/sieve	air/water air/Isopar-M
Wijn	[14]	Weeping	data from literature	air/water air/Isopar-M

In order to make an additional statement about the zones imaged by the new miniaturized test set-up presented here, the tray crossflow is also observed by means of an IR-camera. These measurements are used to highlight significant differences between the test columns and the miniaturized plant, thus simplifying the determination of scaling factors. Table 2 shows an overview of some investigations that have dealt with the detection of the tray crossflow so far. These references will be used to classify and evaluate the data recorded experimentally here.

Table 2: *Small literature overview for investigations of tray crossflow*

Reference		**Method**	**Column/tray**	**Material system**
Schubert	[15]	Wire mesh sensor	$d_i = 0.8$ m/sieve	air/water with conductive tracer
Solari	[16]	Fluorimetric technique	$d_i = 1.22$ m/sieve	air/water
Li	[17]	Infrared camera	$d_i = 1.20$ m/sieve	air/water

Experimental Set-Up

Motivated by the aforementioned requirements the Ruhr University Bochum designed a miniaturized measuring cell. The system has a rectangular shape (0.2 x 0.6 m) and is built up of several transparent section which can be reconfigured in a modular way. The examined trays are flanged between the individual sections. There are options to replace active areas, change weir heights, tray spacing, weeping/entrainment collection system and number of measured trays. The rectangular structure has the advantage of avoiding expansion and constriction of the liquid flow at the outlet of the downcomer as well as at the exit weir. In addition, this design allows measurements with a high-speed camera without distortion. The first test set up is made of transparent PVC in order to initially observe the flow pattern and other variables. After all measurements are completed, a recommendation is made for the smallest possible unit, which can then be set up in a laboratory exhaust. All relevant geometric data is listed in table 3.

Table 3: *Geometric data of the column.*

l [m]	**w** [m]	**h** [m]	n_{Trays} [-]	A_c [m²]	A_{ac} [m²]	l_w [m]	$F_{ac,min/max}$ [$Pa^{0.5}$]		$\dot{u}_{L,min/max}$ [$m^3m^{-2}h^{-1}$]	
0.6	0.2	0.5	2	0.045	0.012	0.2	0.3	2.2	10	40

Experimental Set-Up for hydrodynamic measurements

In order to evaluate the behavior of the fluid dynamics and thus the new setup, many parameters must be measured. Gas and liquid loads, pressure drop, air humidity, stream temperatures (TI1-TI4), weeping, entrainment and two-phase layer height can be measured at the moment. A 3D-model and a schematic process flow diagram are given in Figure 1.

From the storage tank S2 the liquid is pumped (P2) to the first downcomer via the liquid inlet. This is designed as a false downcomer. Via the inlet weir with a height of h = 40 mm, the liquid reaches the second test tray before it leaves the section via the outlet weir with a height of h = 50mm. Both the inlet and outlet weirs are screwed, so that any other combination of heights can be guaranteed. The liquid flows to the first test floor via the second downcomer. After passing the tray it leaves the system using liquid pump P1 through the last downcomer. This one is designed as a closed downcomer to separate weeping from liquid that follows the tray path. Weeped liquid is collected over a chimney tray, which on the one hand collects the weeping (s.f. FR2) and on the other hand provides the gas distribution. Weeping liquid is collected in a small storage tank and recirculated to the container S1. Another special feature is the possibility to divide the collector into several segments, which allow to record the weeping locally. Entrained liquid is also collected at the top of the column. This tray is designed as a sieve tray with small tubes (s.f. FR1). The measured liquid is fed back to the second downcomer.

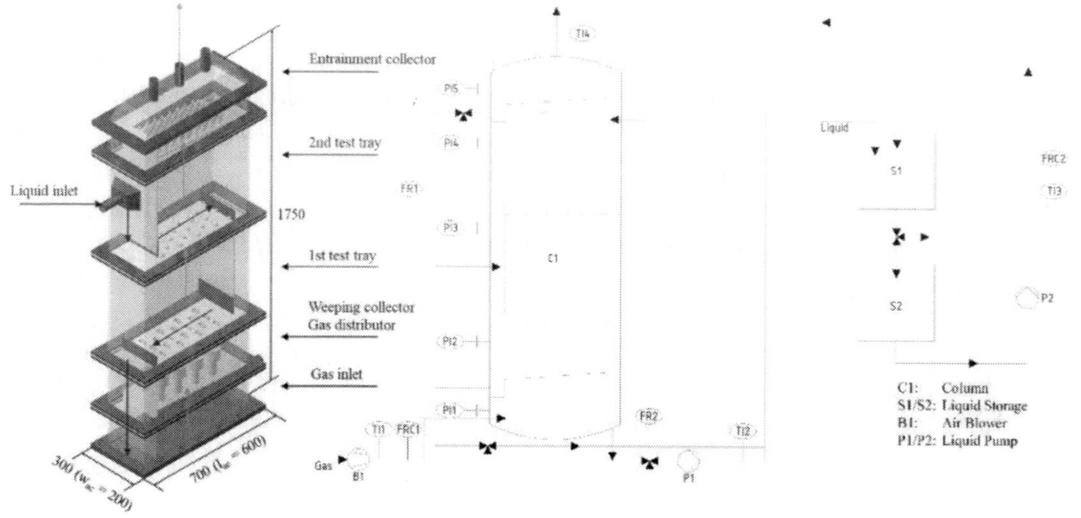

Figure 1: *3D Model and schematic process flow diagram of the Measuring Cell.*

For gas flow ambient air is compressed via blower B1. It is measured by an orifice plate according to DIN EN ISO 5167-2 and supplied to the column via a cap distributor. As aforementioned the gas passes the weeping collector and is equally distributed. The gas passes the two test trays that were installed to realistically represent the hydrodynamics and the interactions between one another. Afterwards the gas passes a demister to remove liquid droplets from the existing gas. The pressure drop is recorded with sensors referring to PI1-PI5. They all measure the differential pressure over the trays. PI5 measures the pressure drop to the ambient. For measurements with the water/air system the gas leaving the column is not recirculated. For measurements with organic substitute systems, recirculation of the gas with a separator is provided due to stripping effects. Since this paper deals in particular with the substance system water/air, the setup is not listed here. In addition, further information on the structure of the individual sections can be found at *Leushacke et. al.* [18]. For the first study, three different types of trays are examined. The data is presented in Table 4.

Table 4: *Geometrical data of the presented tray configurations.*

Tray	ID	φ	d_h	h_{lift}	n	t_{tray}
Sieve	ST6	0.106	10 mm	-	124	6 mm
Sieve	ST2	0.106	10 mm	-	124	2 mm
Fixed valve R-MV (Raschig)	VT	0.106	-	4.5 mm	42	2 mm

Experimental Set-Up for crossflow determination

To determine the crossflow over the tray an IR-camera is used. For these measurements the column needs to be open because PVC is not IR-transparent. In addition, gas loads between $F_{V,ac}$ = 1.2-1.6 $Pa^{0.5}$ and liquid loads between $u_{L,ac}$ = 10 – 40 $m^3m^{-2}h^{-1}$ were measured. For experimental measurements hot water is used as a tracer (T_{HW} = 50°C). As soon as a steady state has been established on the trays, five liters of water are added to the false downcomer over a period of approx. t = 10 s. The camera records data until the initial temperature of the previously defined steady state is reached again. Only the active area is recorded as a section of the image.

Experimental Results

In order to validate that the system presented here can be used to determine fluid dynamics, measurements are made with sieve trays, as these are very well documented in the literature. A comparison with correlations will be consulted to evaluate the system as suitable. These measurements and first measurements with the fixed valve test tray will be presented in the following.

Pressure drops

The measured dry and wet pressure drops for the sieve tray with a thickness of t = 2 mm for the test system air/water at ambient conditions are shown in Figure 2/3. They are compared to with correlations from the literature.

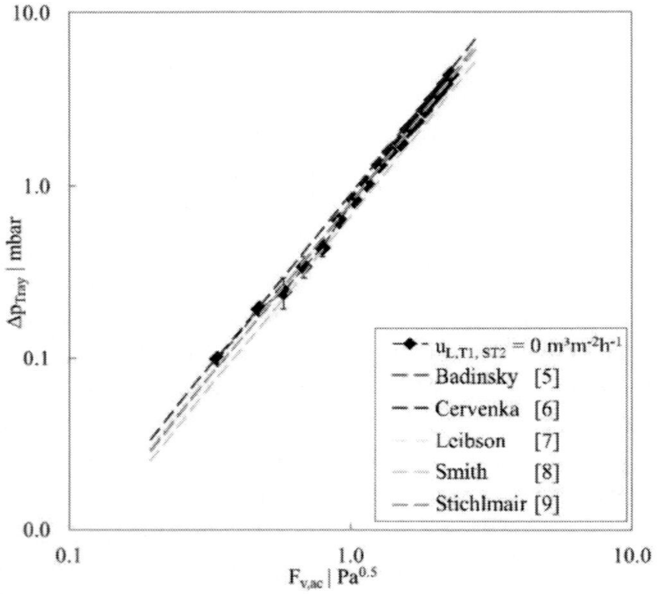

Figure 2: *Dry pressure drop for sieve tray (t = 2 mm).*

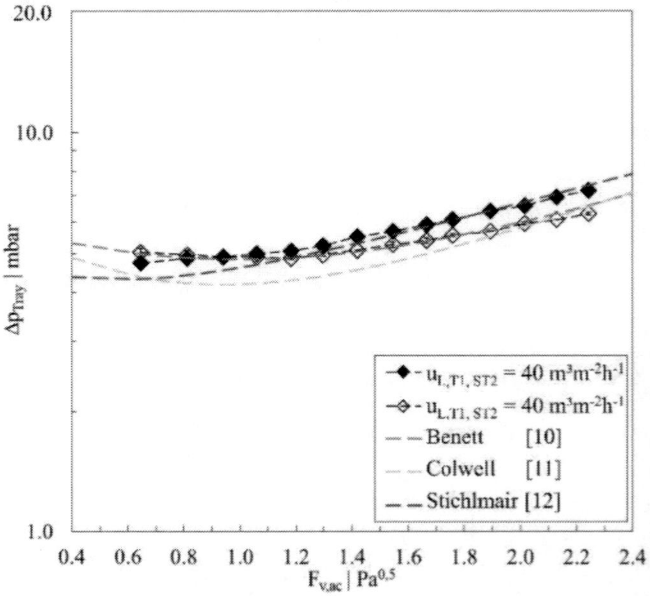

Figure 3: *Wet pressure drop for sieve tray (t = 2mm) for u_L = 40 $m^3 m^{-2}h^{-1}$.*

Figure 2 shows the dry pressure drop over the gas load factor for the first test tray. The pressure drop increases with increasing gas loads. The higher standard deviation for small pressure drops in the range of $F_{V,ac}$ = 0.2 -0.8 $Pa^{0.5}$ are explained by the small absolute values and the logarithmic order. As also described by the correlations, the pressure drops also increase linearly in double logarithmic order. The dry pressure drop can be described very well by the calculations. For the sieve tray with t = 2 mm all the presented correlations fit. Nevertheless, Stichlmair correlation best reproduce the measured values. Figure 3 represents the wet pressure drop over the gas load factor for both test trays and a liquid load of u_L = 40 $m^3m^{-2}h^{-1}$. As expected the wet pressure drop increases with increasing gas load. For $F_{V,ac}$ = 0.6 – 1.0 $Pa^{0.5}$ the pressure drop of test tray 2 is higher than for test tray 1. After this point that behavior changes. This change should not occur due to better distribution of the gas after passing the first test tray, but can be explained by the occurrence of weeping and thus slightly different liquid loads on the trays. As for the dry pressure drop, the behavior of the trays is well reproduced by the correlations. Here, the Benett calculations are found to be the best fit.

On the basis of these two representations, it can be stated that the measuring cell presented here can reproduce the behavior with regard to pressure drops very well. For this reason, a changeover to other trays has taken place, which will be compared with each other afterwards. This comparison is shown in figures 4 and 5.

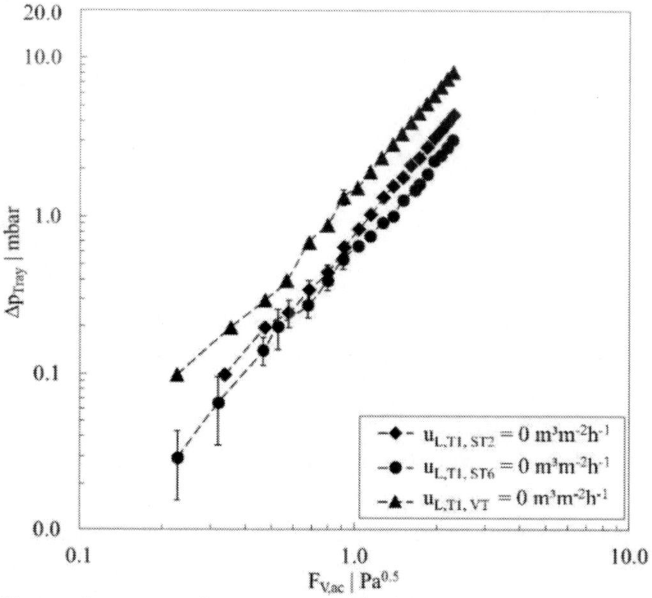

Figure 4: *Comparison of dry pressure drop for all presented trays.*

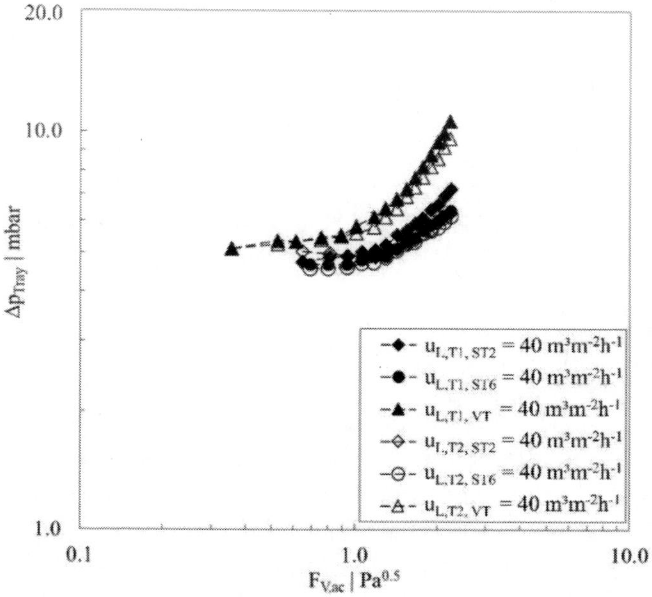

Figure 5: *Comparison of wet pressure drop for all presented trays for $u_L = 40\ m^3 m^{-2} h^{-1}$.*

Figure 4 shows the dry pressure drop for all trays presented in Table 4. As expected, the valve tray has the largest pressure drop followed by the sieve tray with a thickness of $t = 2$ mm and the sieve tray with $t = 6$ mm. This result also shows that the system can be used to determine the fluid dynamics of tray columns. To round out the plot, Figure 5 shows the wet pressure drop for all trays at a liquid loading of $u_L = 40\ m^3m^{-2}h^{-1}$. Again, the pressure drop for the second test tray is always lower for all trays. Since the weeping is only pumped back into the storage tank, liquid weeping through only reaches the first test tray, which may explain the higher pressure drops. Nevertheless, the classical trend of the pressure losses for all trays is shown.

Weeping

Figure 6 shows the comparison between the measured weeping rate for the sieve tray with a thickness of $t = 6$ mm and the models of Wijn and Lockett for a liquid load of $u_L = 40\ m^3m^{-2}h^{-1}$. Neither of the two models shown here can accurately replicate weeping. Although the course for low gas loads - and thus the important part - can be reproduced, the course for high gas loads does not match. Other investigations could also show that further research is necessary for an appropriate representation of the measured values [19]. However, it cannot be excluded that weeping in the rectangular system cannot be properly reproduced. Also, a comparison of the weeping rates for the different trays does not allow a clear statement here (c.f. Figure 7).

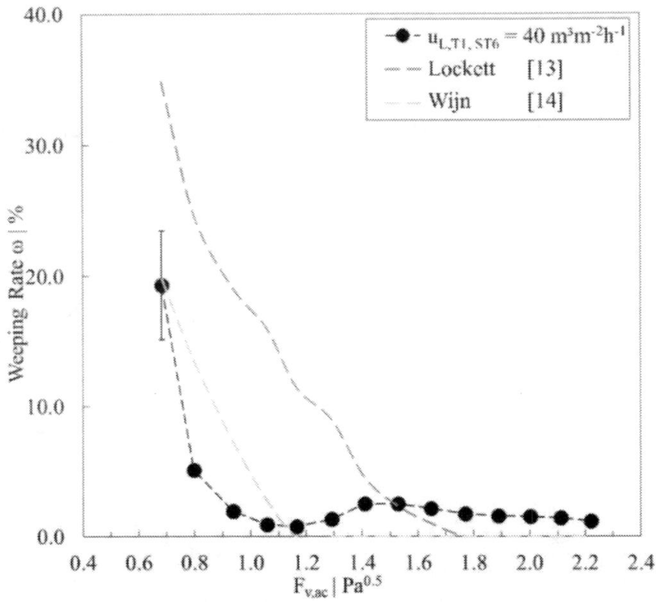

Figure 6: *Weepingrate for sieve tray (t = 6 mm) for $u_L = 40\ m^3 m^{-2}h^{-1}$.*

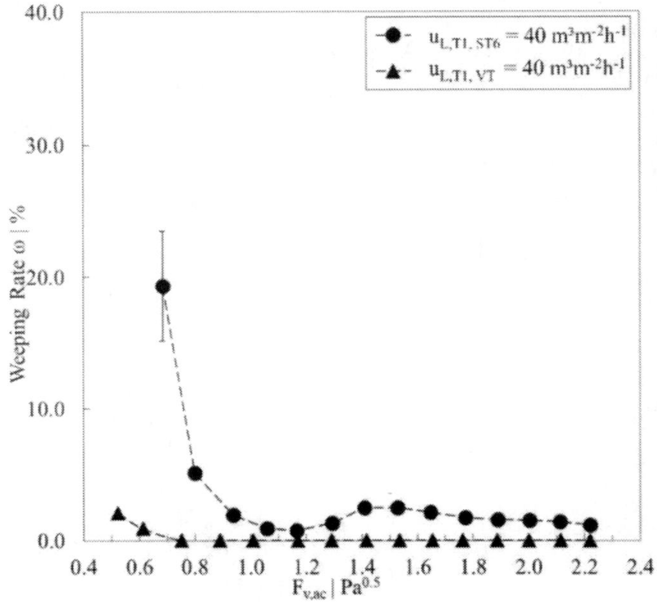

Figure 7: *Comparison of weepingrate for all presented trays for $u_L = 40\ m^3 m^{-2}h^{-1}$.*

As expected, the valve plate shows a better weeping behavior. A comparison of other liquid loadings with literature data shows that the critical point for weeping with the miniaturized measuring cell occurs significantly later (not shown here). One reason for this may be the nature of the tray crossflow. The ideal flow should be reproduced in the miniature measuring cell. Gas and liquid are almost equally distributed, which does not occur in industrial or round columns. Accordingly, zones that are not well gassed, which are decisive for weeping, do not occur in the rectangular column, which explains lower weeping. This behavior is represented in Figure 8.

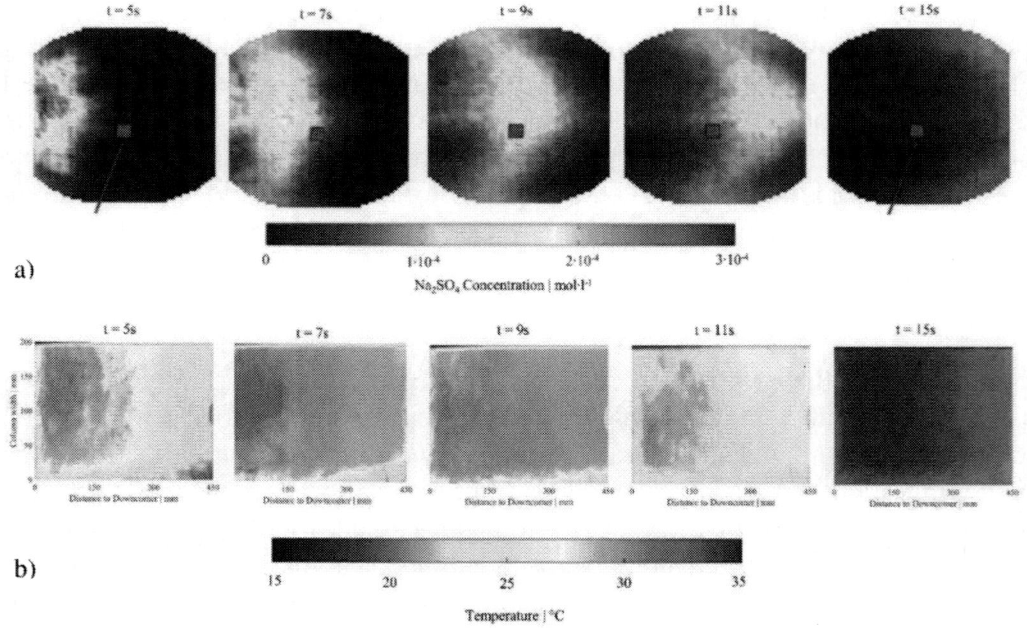

Figure 8: *a) Tracer passage for $u_L = 6\ m^3m^{-2}h^{-1}$ and $F_V = 0.72\ Pa^{0.5}$[15] and b) Crossflow measurements with IR-camera in rectangular system for $u_L = 40\ m^3m^{-2}h^{-1}$ and $F_{V,ac} = 1.6\ Pa^{0.5}$.*

A comparison of the tray crossflow between a round column and the rectangular column presented here. *Schubert et al.* investigated the tray crossflow in an ID800 column using tracer measurements and a wire mesh sensor. These measurements are shown in the upper part. The lower part of the figure shows in comparison the measurements with the IR camera on the column presented here. It can be observed very well that the tracer in the round column spreads over the cross section to different extents, whereas clear tracer fronts can be seen in the rectangular column. These fronts show uniform distribution over the tray.

Conclusion

One of the prioritized topics in the design process of tray columns is the determination of optimal flow conditions and to provide experimental data. To avoid high costs when testing new trays in large laboratory columns, a rectangular miniaturized test rig was developed at Ruhr University Bochum. Test measurements with sieve trays to evaluate the test system were carried out and it could be shown that the pressure drops are mapped reliably. A comparison of weeping measurements showed that such phenomena cannot be reliably represented in rectangular columns so far due to the crossflow. Measurements with IR camera show that the tray crossflow differs significantly between round and rectangular columns. Further investigations are necessary to better show the behavior. In addition, the investigation of further fluid dynamic quantities should be in the foreground. In the future, the test rig will also be used to reliably measure organic substitute systems. Subsequently new tools and methods to precisely predict the behavior of columns at their operating limits and in between will be developed.

Acknowledgement

The project is supported by the Federal Ministry for Economic Affairs and Climate Action, the AiF (IGF project-number: 20835 BG2) and IGF. The authors gratefully acknowledge this support as well as the cooperation with the project partners of Helmholtz-Zentrum Dresden-Rossendorf and Technical University Munich.

Nomenclature

A	area, [m²]
A_{ac}	active area, [m²]
A_c	column area, [m²]
d_h	hole diameter, [mm]
d_i	inner diameter, [m]
$F_{v,ac}$	gas load factor active area, [$Pa^{0.5}$]
h	tray spacing, [m]
h_{clear}	clear liquid height
h_{lift}	valve lift height, [mm]
ID	Identifier
l	length column, [m]
l_w	length weir, [m]
n_h	number of holes
n_{Trays}	Number of trays
p_{dry}	dry pressure drop, [mbar]
p_{total}	total pressure drop, [mbar]
t	time, [s]
t_{Tray}	tray thickness, [mm]
T	temperature, [°C]
T_{HW}	hot water temperature, [°C]
$u_{L,ac}$	liquid load active area, [$m^3m^{-2}h^{-1}$]
w	width column, [m]
Δp_{tray}	pressure drop over tray, [mbar]
φ	relative hole area

References

1. Stichlmair, J.; Klein, H.; Rehfeldt, S.: „Distillation: Principles and Practice.", *John Wiley & Sons,* 2021
2. Repke, J.-U.; Kohrt, M.; Wozny, G.: „Untersuchungen mehrphasiger Strömungen an einfachen geneigten Platten als potenzielles Auslegungswerkzeug für technische Trennapparate", *Chemie Ingenieur Technik, 83, 7,* 2011
3. Stoter, F.; Olujić, Ž.; de Grauuw, J.: „Modelling and measurement of gas flow distribution in corrugated sheet structured packings", *The Chemical Engineering Journal, 53, pp. 55-66,* 1993
4. Plate, D.; Brinkmann, J. H.; Merkel, A.; Gruenewald, M.: „Stofftransportmessungen in einer Messzelle zum Scale-Up von Packungskolonnen", *Jahrestreffen ProcessNet Fluidverfahrenstechnik,* 2021
5. Badínský, M.; Rylek, M.: „Dry pressure dops of Ripple and Kittel plates", *Collection of Czechoslovak Chemical Communications 34 (10), pp. 3156-3159,* (1969)
6. Červenka, J.; Kolář, V.: „Hydrodynamics of plate columns. VIII. The dry-plate pressure drop of sieve-plate separating columns", *Collection of Czechoslovak Chemical Communications 38, pp. 2891–2897,* (1973)
7. Leibson, I.; Kelley, R. E.; Bullington, L. A.: „How to Design Perforated Trays", *Petrol. Ref. 36 (2), pp.127-133,* (1957)
8. Smith, P. L.; van Winkle, M.: „Discharge Coefficients through Perforated Plates at Reynolds Numbers of 400 to 3.000", *AIChE Journal 4 (3); pp. 266-268,* (1958)
9. Stichlmair, J.: „Dimensionierung des Gas/Flüssigkeit-Kontaktapparates Bodenkolonne, Teil 2", *Chemie Ingenieur Technik 50 (5), pp. 383-387,* (1978)
10. Bennett, D. L.; Agrawal, R.; Cook, P. J.: „New pressure drop correlation for sieve tray distillation columns", *AIChE Journal 29 (3); pp. 434–442,* (1983)
11. Colwell, C. J.: „Clear liquid height and froth density on sieve trays", *Industrial & Engineering Chemistry Process Design and Development 20 (2, pp. 298–307),* (1981)
12. Zuiderweg, F. J.: „Sieve trays-A view on the state of the art", *Chem. Eng. Science 37 (10), S. 1141-1464* (1982)
13. Lockett, M. J.; Banik, S.: „Weeping from Sieve Trays", *Ind. Eng. Chem. Process Des. Dev. 25 (2),* (1986)
14. Wijn, E. F.: „On the lower operating range of sieve and valve trays", *Chemical Engineering Journal 70, pp. 143-155,* (1998)
15. Schubert, M.; Piechotta, M.; Beyer, M.; Schleicher, E.; Hampel, U.; Paschold, J.: „An imaging technique for characterization of fluid flow pattern on industrial-scale column sieve trays", *Chemical Engineering Research and Design III, pp. 138-146;* (2016)
16. Solari, R. B.; Bell, R. I.: „Fluid flow patterns and velocity distribution on commercial-scale sieve trays", *AIChE J. 32 (4), pp. 640-649,* (1986)
17. Li, Y.; Wang, L.; Yao, K.: „New technique for measuring fluid flow patterns on a multiple downcomer tray", *Ind. Eng. Chem. Res. 46, pp. 2892 – 2897,* (2007)
18. Leushacke, W.; Grünewald, M.: „Konzeption einer miniaturisierten Messzelle zur Auslegung von Hochleistungsböden", *Poster, Jahrestreffen ProcessNet Fluidverfahrenstechnik und Wärme- und Stoffübertragung, online,* (2021)
19. Engel, F. S.; Klein, H.; Rehfeldt, S.: „Hydraulic Investigations of the Operating Limits of High Performance Trays", *12th international conference D&A,* (2022)

157a Refinery to Petrochemical Integration Using Dividing Wall Column Technology

Manish Bhargava, Amit Kanda

With lower gasoline demand, an integration between a refinery and petrochemical complex has become more lucrative. Efforts are being made in identifying synergies and optimizing them for operational and economic gains where refinery and petrochemical complex exchange many streams such as feedstocks, byproducts and utilities.

Dividing wall columns offer an attractive option for an integration between refinery and petrochemical complex. The paper presents the following case studies of using dividing wall column technology for this integration:

- Meet new specifications of 1 ppm CS2 in petrochemical naphtha
- Using byproducts from isomerization unit as petrochemical feedstock
- Revamp existing naphtha splitter to maximize aromatic feed stock for petrochemical complex.

157b Dividing Wall Columns - Novel Applications with Case Studies

Robin Schulz, Anand Vennavelli, Neil Sandford

Reduction of CO2 emissions related to distillation applications necessarily involves reducing the energy demand. Dividing wall column technology can play an important role toward this goal as it can achieve significant energy savings of up to 50% compared to conventional designs.

Dividing wall columns were first described in the first half of the 20th century in two U.S. patents [1,2]. The concept combines the heat-integrated connection of two columns in one apparatus to save energy. Later, thermodynamic and simulation improvements [3,4,5] allowed for a simple and fast simulation of a dividing wall column.

The first commercially used dividing wall column was built and put into operation in the mid-1980s. The column was equipped with internals by Julius Montz GmbH. The combination of thermodynamical knowledge of the operating company and the hydraulic knowledge of column internals by Montz was a decisive contribution to the proper usability of a dividing wall column.

A good understanding of the hydraulic behavior is critical for a proper design of a dividing wall column. For instance, improper distribution of the vapor flow below the dividing wall to each side of the wall will result in the actual separation being significantly different from the simulation results.

Initially, the challenges of dividing wall column design and unfounded concerns about the difficulty of controlling such columns led to the fact that only a few large petrochemical and chemical companies were willing to design, build and operate dividing wall columns.

Even today, opinion persists that dividing wall columns are only a marginal phenomenon in distillation and that even the mechanical challenges are too great to bring dividing wall columns to a wider range of applications. To counter this view, we will present two case studies featuring dividing wall columns, one containing packing and the other trays. These case studies will show that for almost every theoretical obstacle to the design of a dividing wall column, a mechanical solution already exists that has also been tested and proven on an industrial scale.

In conclusion, the recommendation to all operators of distillation plants is to realize the potential energy savings of up to 50% compared to a conventional column system, and to also consider the use of dividing wall columns for existing processes.

References

1. D. A. Monroe, Fractionating Apparatus and Method of Fractionation, 2,134,882, United States Patent Office (1935).
2. R. O. Wright, Fractionation Apparatus, 2,471,134, United States Patent Office (1946).
3. F. B. Petlyuk, V. M. Platonov, D.M. Slavinskii, Thermodynamically optimal method for separating multicomponent mixtures, Int. Chem. Eng. 5, 3 (1965), 555-561.
4. I. Dejanovic, Lj. Matijasevic and Z. Olujic, Dividing wall column – A breakthrough towards sustainable distilling, Chem. Eng. Process. 49 (2010), 559-580.
5. I. J. Halvorsen and S. Skogestad, Minimum Energy Consumption in Multicomponent Distillation. 2. Three-Product Petlyuk Arrangements, Ind. Eng. Chem. Res. 42 (2003), 605-615

157d Network of Dividing Wall Columns in a Grassroots Chemical Plant

Amit Kanda, Manish Bhargava

When plot space is a critical limiting factor dictating the feasibility of a new, one-of-a-kind chemical plant, combining multiple columns into a single divided-wall-column does not stay only as an option, but becomes a necessity. In such cases, even one dividing wall column is not adequate, and companies need a series of inter-connected dividing wall columns (DWC's) to minimize plot space requirements in order to be able to implement the new process.

Added to it, the complexities of multiple feed and operating cases, reactor recycle streams, presence of water and other trace impurities, and requirement of multiple high purity products – the challenge of implementing DWCs becomes many-fold. Of course, energy efficiency, equipment count reduction and utility savings are the obvious by-products of implementing the DWC's.

The paper discusses the following:

1. Combining eight conventional columns into a series of four divided-wall-columns
2. Multiple instances of dual-divided-wall columns in the scheme.
3. Significant equipment count reduction
4. Energy savings of approx. 30%

157e Revamping a Conventional Naphtha Splitter to a Dividing Wall Column at a Major Southeast Refinery

Giuseppe Mosca, Andrei Cimpeanu, Brad Fleming

The existing Naphtha Splitter is a conventional side draw column; it splits the feed coming from the bottom of an FCC Unit De-Pentanizer into three streams: Naphtha Splitter Light Cut, Naphtha Splitter Middle Cut and Naphtha Splitter Heavy cut.

The heat duty to the column is provided by two reboilers: Upper and Lower reboiler, while the top condensation system provides the cooling duty to generate the required internal reflux. The feed is pre-heated at the pre-heat exchanger by mean of the side cut draw off before entering the middle section of the tower.

A poor fractionation was occurring between the adjacent cuts resulting in a huge overlap, particularly between the Middle and the Heavy Cuts.

To overcome these limitations, it was decided to convert the tower into a Dividing Wall Column.

To successfully convert a conventional Naphtha Splitter to a Dividing Wall Column, a deep knowledge of the process is a must, but certainly not enough. It is also crucial to fully comprehend the equipment being used, their mechanisms of operation, their limitations, and capabilities. This includes, among others, the mass transfer components being used in the baffled sections of the tower: half-moon shape fractionation trays, and half-moon shape packed beds, and the associated distributors.

This paper will describe the revamp of a conventional Naphtha Splitter into a Dividing wall column from the conceptual design to the commissioning of the revamped equipment and the operating data after the revamp

25d Improving Vacuum Tower Asphalt Quality

Andrew Sloley

Improving Vacuum Tower Asphalt Quality

Andrew W. Sloley

The Distillation Group, Inc.

Bellingham, WA, USA

Asphalt is one of the lower value products from a refinery. However, a market exists for road paving, roofing, and other asphalt types. For refineries that lack bottoms conversion capability being able to meet asphalt specifications may be critical to overall refinery economics. This case study looks at a refiner with limited residual conversion capacity. Vacuum residue rom the crude vacuum unit goes to the asphalt and road oils market.

During their typical three-year run, production of road asphalts becomes more difficult during each successive summer. Finally, production of specific asphalt grades requires reduced crude rates. The reduction in crude runs directly reduced profits by limiting gasoline production during peak demand months. The objective of the work was to allow on-specification asphalt production at full crude rates during the summer.

Diagnosing a Premature Flood Near the Diesel Draw of an Atmospheric Crude Tower

By

Daniel Hussman, Gord Bruce, Abdullah Abufara, and Komi Chandi, Parkland Refining (BC) Ltd., Burnaby, BC, Canada

and

Henry Z. Kister, Fluor, Aliso Viejo, CA, USA

Presented at the Distillation Symposium, AIChE Spring Meeting, Houston, Texas, March 12-16, 2023

DIAGNOSING A PREMATURE FLOOD NEAR THE DIESEL DRAW OF AN ATMOSPHERIC CRUDE COLUMN

By Daniel Hussman, Gord Bruce, Abdullah Abufara, Komi Chandi, Parkland Refining (B.C.) Ltd, Burnaby, BC, Canada, and Henry Z. Kister, Fluor. Aliso Viejo, CA, USA

The most recent turnaround for this column was in 2020, at which time there were no abnormal findings near the diesel draw and in the wash section below. The planned mechanical work was completed, wash section 2-pass sieve trays 8-10 in the tower were replaced by in-kind, but the tray material was changed from 410 SS to Carbon Steel. Then the column operated at planned rates for 18 months with no issues. In October 2021, an adjacent unit was offline for a planned turnaround, and this atmospheric crude column was drained, purged and safe parked for the duration of the planned event.

Following the October 2021 outage, the column began to struggle in maintaining planned rates due to increased pressure drop across the bottom 2/3 of the column. Over the next 7 months, the pressure drop would rise in a stepwise fashion resulting in rate cuts upwards of 5-6 MBPD. The tower appeared to be flooding at higher crude rates, causing sudden loss of over-flash and diesel draw (level in the side stripper). The flood diagnosis was confirmed by gamma scans. The "normal operation" scan showed flooding on tray 10, right above the diesel draw tray. In the "normal operation" scan, the flood did not propagate further up. The "flooded" scan showed the flood propagating several trays up the tower.

Several theories, requiring different remedial actions, were proposed. The theories were evaluated by thorough checking against plant data together with applying quantitative analysis to critically evaluate gamma scans. Based on this evaluation, the team concluded that entering the tower to clear mechanical restriction/damage or fouling is the best path forward. A detailed cost-benefit analysis was completed a few different ways to reflect the fluidity of the supply and retail markets as global tensions peaked. Upon entry, the active areas on Trays 10 and 9 (diesel draw) were found heavily fouled, to the point where the sieve holes were no longer visible. The fouling thickness varied across the diameter of the column and could be easily removed with a metal scraper. The deposits were clay-like and sandy. The downcomers were also noted as relatively clean. Trays above and below this zone were found to be quite clean given the service.

Our paper describes the troubleshooting investigation and demonstrates the importance of closely checking theories against plant data for diagnosing fractionator problems.

PROCESS DESCRIPTION AND PROBLEM DEFINITION

Figure 1 is a schematic of the crude tower. The rates and temperatures shown are typical values for the 18 months before the October 2021 outage. The tower differential pressure, measured between the flash zone and the vapor space above tray 24 was steady, typically 2.3 – 2.5 psi, or about 0.13 psi/tray, a little on the high side, but within the range of normal operation. The crude rate was maxed out (30MBPD), diesel pumparound was high (5,700BPD), diesel side stripper level steady, overflash steady at about 500 bpd.

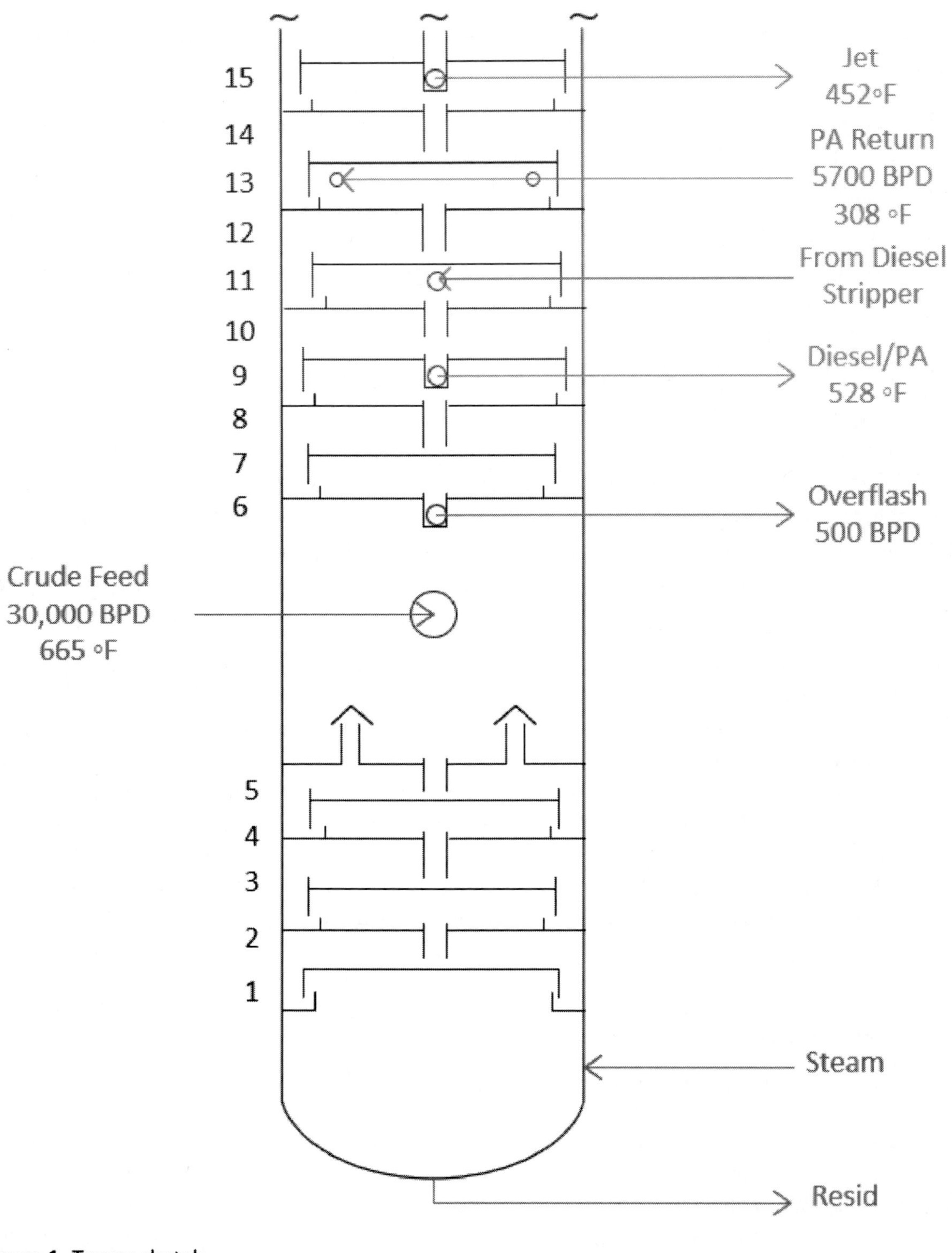

Figure 1. Tower sketch

After the October 2021 turnaround the tower dP became much more erratic and feed rates had to be limited to 29,000 to prevent dP excursions. The diesel pumparound rate was reduced to about 4500 BPD with colder outlet of about 225°F.

There were two temperature indicators in the vapor space between trays 8 and 9. Historically, these two temperature indicators read the same. Following the October 2021 outage, the one on the northeast started reading 15°F higher than the one on the southwest. There were two other temperature indicators in the vapor space between trays 6 and 7, and they kept reading the same temperatures.

Things took turn to the worst following another outage in December 2021. The dP became more erratic, often rising to about 3.5-4 psi. When the dP rose overflash was often lost. Liquid appeared to be accumulating above the diesel draw with a loss of diesel product and of wash to the lower trays. The high dP was often accompanied by some diesel ending in the jet draw from tray 16. The accumulation appeared sudden, erratic and unpredictable. Pulling out more diesel brought the dP down, but overflash would not re-establish until the coil outlet temperature (normally about 650-655°F) was lowered by about 5°F. The strong dependence on the coil outlet temperature suggests the flood was strongly sensitive to the vapor loading.

The flood typically appeared as a sudden partial loss of liquid flow into the diesel stripper and loss of all liquid flow down the tower from tray 9 into the trays below. The sudden and unpredictable loss of liquid flow at the diesel tray was a major operation problem.

The frequency and severity of the flooding above the diesel draw tray appeared to increase at higher crude charge rates and high diesel pumparound flow rates. This forced the refinery to gradually cut charge rates, from 28,000 BPD after the December outage to 25,000 BPD just before the tower was shut down in July 2022. Diesel pumparound flow rates were further reduced to about 2,200 BPD.

TROUBLESHOOTING AND TESTING

The diesel pumparound temperatures were lower than the normal operations due to a couple of the feed pre-heat exchangers being fouled. So, the early theory was that the sub-cooled pumparound return to the column reduced distillation efficiency by converting the mass transfer zone to a heat transfer zone and increasing the liquid load between Jet and Diesel draws (higher internal reflux). The exchangers were cleaned or were partially bypassed to heat up the internal reflux stream, but the column performance issues continued. Also, In the past the crude tower was able to run at both high crude charge rates and diesel pumparound flow rates while the diesel pumparound return temperature was low. Therefore, this ruled out the low diesel pumparound temperature as the root cause of the problem. However, one test gave evidence that it could be a contributing factor. At this test an increase in the pumparound flow rate caused a dP spike, inducing flood.

The tower was gamma-scanned in the flooded and "normal operation" conditions. The scans were conducted on January 25th (flooded) and January 26th ("normal operation"). During both scans, the crude feed rate was 28,000 BPD and the diesel pumparound flow rate 3,800 BPD. The key parameter changed was the coil outlet temperature. At the flooded scan the flash zone temperature was 653°F, lowered to 646°F to give the "normal operation" scan. The pressure drop at the flooded scan ranged from 2.8 to 3.7 psi, compared to 2.3 psi and stable in the "normal operation" scan.

The flooded scans (Figure 2, blue and red pens) showed flooding propagating from tray 10 upwards all the way to trays 17-19 (Figure 2 was cut off at tray 15, but the scans continued all the way up the tower). In the flooded region (tray 10 and up), the scans show little difference between the northeast (NE) and southwest (SW) chords. On the unflooded trays below (6 through 9), the SW side (red pen) showed taller and denser froths than the NE side (blue pen). The NE sides of trays 6, 7, and possibly also 9 looked dry while the SW sides looked normal and not heavily loaded.

The "normal operation" scan showed flooding on tray 10 only. Trays 11 and higher were not heavily loaded, with little difference between the NE and SW chords. This verifies that the flooding is due to a restriction near tray 10 and not to tray design issues on tray 11 or higher up. The taller and denser froths observed on the SW of trays 6-9 in the flooded scan persisted on tray 9 in the "normal operation" scan but became much less pronounced on trays 6-8. Only tray 6 in the "normal operation" scan NE side approached drying.

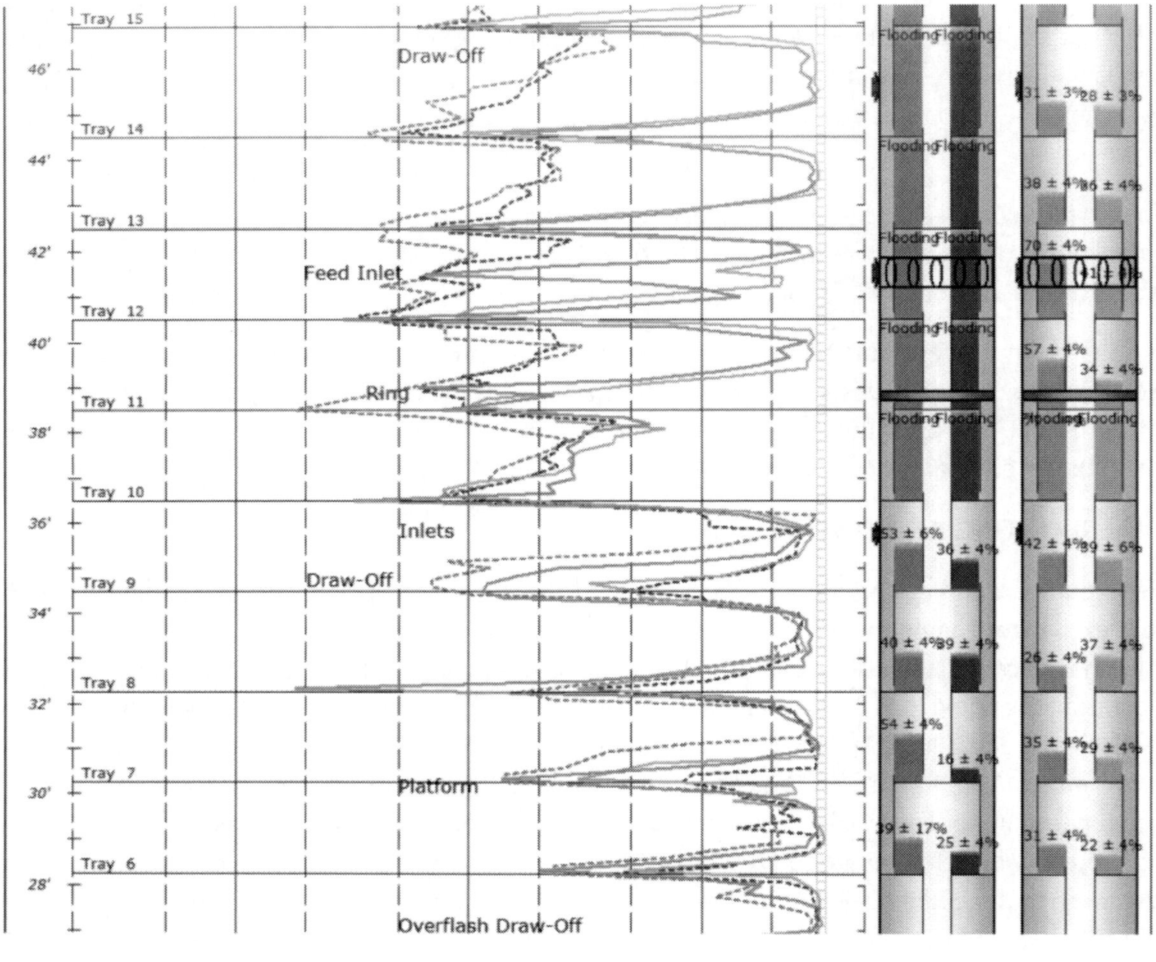

Figure 2. Active area gamma scans under flooded and "normal operation" (pink and gold pens) conditions.

Legend for Figure 2:

Blue pen, flooded scan, northeast (NE) chord

Red pen, flooded scan, southwest (SW) chord

Gold pen, "normal operation", NE chord

Pink pen, "normal operation", SW chord

Center downcomer scans of trays 6-15 (Figure 3) were shot two weeks after the active area scans. The intention was to catch the condition at which the downcomers would begin to fill up. During both scans, the crude feed rate was 28,400 BPD and the flash zone temperature was 651°F, similar to those in the flooded active area scans. The diesel pumparound flow rate was a lot lower, 2700 BPD compared to 3800 BPD in the active area scans, intended to unload the downcomers. For the "more loaded" scan (blue pen) the pumparound return temperature was similar to that during the active area scans, 211°F. For the "less loaded" scan (red pen) this temperature was raised to 244°F, to further unload the downcomer. The pressure drop at both tests was about 2.4 psi and stable.

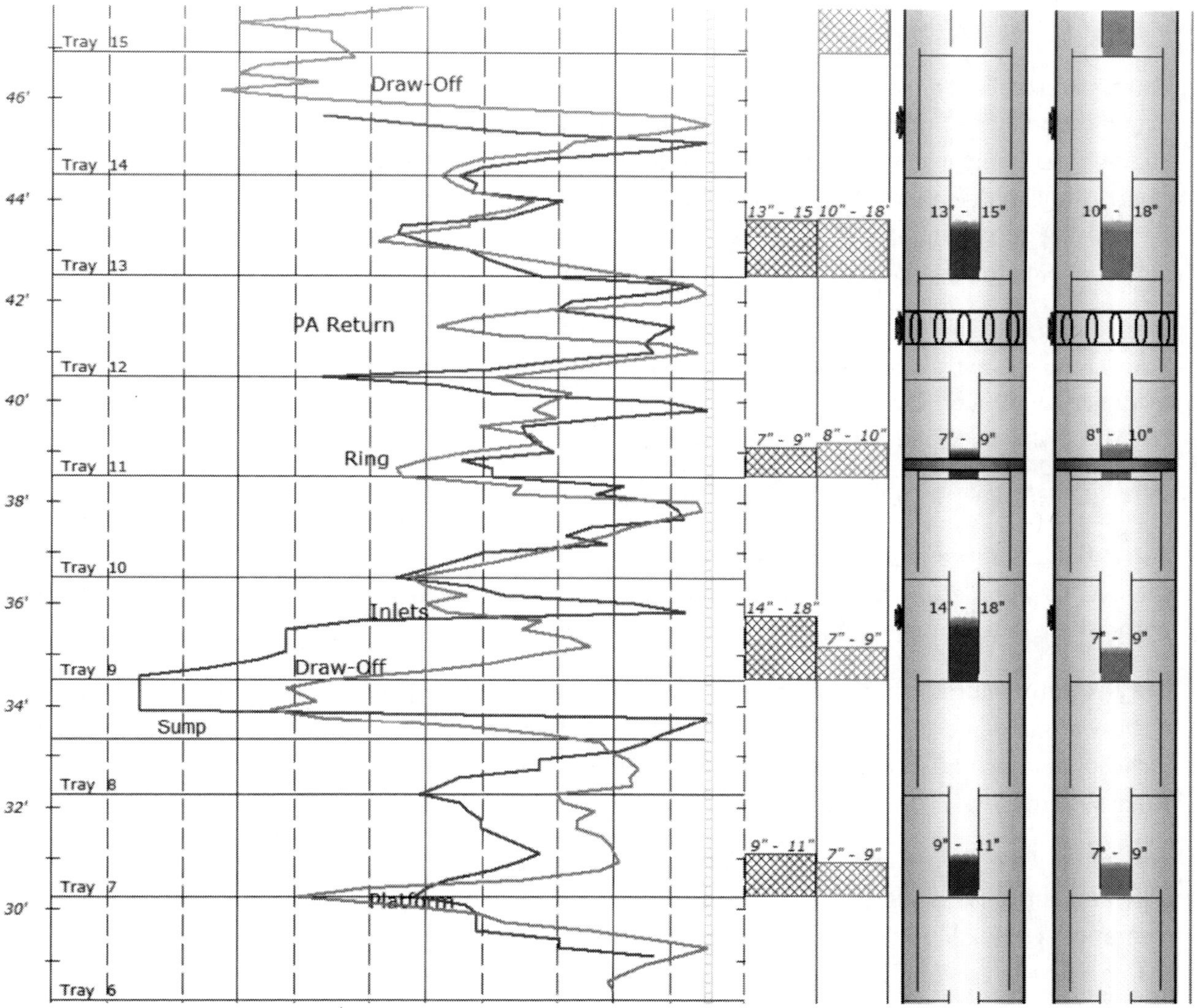

Figure 3. Center downcomer gamma scans for the "more loaded" (blue pen) and "less loaded" (red pen) tests.

Figure 3 shows no flooding in the tower, as confirmed by the low dP. The active area above the center downcomer on tray 10 was close to flood, but the active areas above the center downcomers of trays 12 and 14 were not heavily loaded. This again verifies that the first active area to flood is that above tray 10.

There was a large difference between the two scans in the center downcomer from tray 10. The downcomer froth height in the more loaded scan was about double that in the less loaded scan and approached the top of the downcomer. The froth in the more loaded scan was also much denser than in the less loaded scan. In contrast, the loading difference had little effect on the froth heights and froth density in the tray 12 downcomer (which also serves the pumparound) and its froth did not approach the tray above. This strongly supports a restriction in the tray 10 downcomer, or plugging on tray 10, or both.

There was also a significant difference in the downcomer from tray 8. The downcomer appeared normal in the more loaded scan but headed towards drying out in the less loaded scan.

The tests established that the flood is both vapor-sensitive and liquid-sensitive. The vapor sensitivity was established by the finding that raising the flash zone temperature by as little as 5°F, while keeping the charge rate, pumparound rate, diesel draw rate, and pumparound return temperature constant, induced flood. The liquid sensitivity was established by the froth height and density in the tray 10 downcomer flood rising due to an increase in diesel pumparound rates or lowering pumparound return temperature while keeping the charge rates and flash zone temperature constant.

These scans established that the tower bottleneck is a downcomer backup flood initiating in the downcomer from tray 10 and then building up and flooding tray 10 and above. The scans also provided evidence for uneven drying up of some of the wash section trays when the liquid rate below tray 9 drops off. The drying occurred preferentially on the north.

INITIAL THEORY: BLOCKAGE OF THE TRAY 10 DOWNCOMER

The initial theory was that a side panel in the downcomer from tray 10 has come loose, partially blocking the overflow of diesel out of the diesel draw trough, possibly allowing vapor into the draw box. If vapor entered the downcomer, it would impede liquid flow into the draw box from the tray above.

Tray Layout Figure 4 shows the plan and elevation of trays 6-9. The trays were 10.5 ft ID, 2-pass sieve trays with multi-chordal downcomers. The center and side downcomers were straight, approximately 8.5 inches and 7 inches wide, respectively. The holes were ½" diameter. The hole area was about 12.4% of the tray active area.

The outlet weirs were picketed, with the top of the pickets 14.5" above the tray floor. The weir heights between the pickets were 2.5". The pickets occupied about 86% of the weir lengths. Downcomer clearances were 2.5". There were 3" inlet weirs located 2⅞" away from the downcomer walls.

The floor of the center sump below tray 9 was sloped, ranging from a depth of 8" at the end opposite the draw nozzle to a depth of 16.5" at the nozzle end, with the 6" draw nozzle flush with the bottom of the sump. Beneath the deep end of the sump the picket-fence weirs were shortened to permit a 2-4" gap between the sump and the top of the picket fence weirs.

Strengths and weaknesses of the downcomer blockage theory This theory would readily explain the initiation of flood on tray 10 as seen by the gamma scans. The theory also provided an excellent

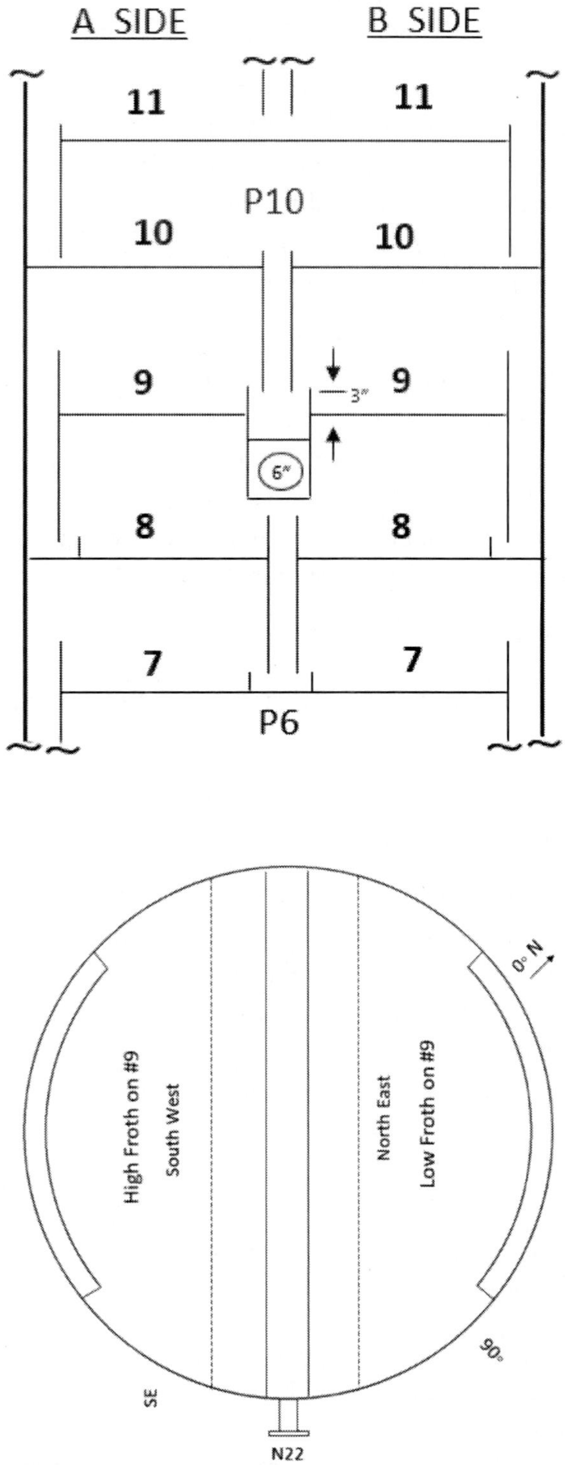

Figure 4. Elevation and plan, trays 6-9.

explanation to the erratic behavior of the flood. Either a liquid buildup step followed by a dump step, or vapor entering the downcomer, will explain the sudden and unpredictable loss of liquid flow at the diesel draw and below.

One issue with this theory is the low velocity in the tray 10 downcomer, estimated at about 0.1 ft/s (including product, reflux, and pumparound), compared to typical maximum design velocities of 0.4-0.5 ft/s (7) for non-foaming systems. The panel blocking the downcomer needed to be large and lodged at an orientation that would block at least 60-80% of the downcomer. Alternatively, the gap that would allow the vapor in needed to be quite large, at least 2 ft^2. Further, the downcomer gamma scans did not show any evidence of vapor breaking into the center downcomer from tray 10.

Another issue with this theory was its inability to explain the maldistribution of liquid below the diesel draw that was observed in the gamma scans. It may be argued that the dislodged panel impedes flow to one side of the tray more than the other. The problem with this explanation is that there are inlet weirs that would equalize the flow out of the inlet weir sump. In addition, due to the large downcomer clearances, the friction liquid head loss at the downcomer apron was very small, about 0.005" (assuming no blockage), so even a significant blockage would not induce maldistribution to the liquid descending to the lower trays. This started to look like the major argument against the downcomer blockage theory.

FOCUS ON MALDISTRIBUTION BETWEEN PANELS

To correctly evaluate the tray 10 downcomer blockage theory, it was important to critically review the evidence for and possible cause of this maldistribution.

Quantitative analysis of gamma scans is an invaluable tool for evaluating maldistribution between panels (1). We therefore applied it for this troubleshooting endeavor.

We calculated froth heights for trays 7 and 9 using the Harrison method (2), with which we have had excellent experience, and which is often considered the most accurate method for froth height determination (1). The results are in Table 1. Also included in the Table are indices for froth density. The index we used was the minimum transmission count of the peak. The lower the value the denser the froth. This is a shortcut method. An accurate evaluation involves integration of the peak density along the peak height (3). This tedious and time-consuming method was not justified here due to time and budget constraints.

TABLE 1. RESULTS OF QUANTITATIVE ANALYSIS OF THE TRAYS 7 AND 9 GAMMA SCANS

TRAY	"normal operation"	"normal operation"	Flooded	Flooded
	NE	SW	NE	SW
7 Froth height, in	8	11	9	18
7 Density, counts	400	200	900	170
9 Froth height, in	12.5	16	13.5	18
9 Density, counts	500	120	400	50

Table 1 dispels any doubts that for both trays 7 and 9 the froth heights were taller, and the froths were denser on the SW on both tests. The density counts show that the NE side of Tray 7 was almost dry in the flooded scan. Tray 9 was also low on liquid. In the "normal operation" scan trays 7 and 9 had more liquid on the NE side than in the flooded scan, but significantly less liquid than on the SW. On the even-numbered trays 6 and 8, the densities again were higher on the SW than on the NE. In the flooded scan the froth heights were also taller on the SW than on the NE. In the "normal operation" scan the difference in froth height between NE and SW were about 3 inches, growing to 6-9 inches in the flooded scan. This analysis established that on trays 6-9, especially the odd numbered, there was a significantly larger quantity of liquid descending on the SW than on the NE.

This maldistribution observation ties in with the plant observation that the temperature indicator on the NE read 15°F hotter than the temperature indicator on the SW. Together, they verified maldistribution between the SW and NE panels, with more liquid descending on the SW.

An earlier article (4) describes a post-mortem analysis of an uneven loss of valve floats from an upper fractionation tray in another crude tower. We applied the Fluor maldistribution model (5) in an endeavor to reconcile the pattern of lost valve floats to the gamma scan results. Case 3 in Ref 4 is relevant (Figure 5). It examines the distribution pattern of vapor and liquid on the trays beneath Tray 4 that had 90% valve float loss on side A and no valve float losses on side B. The modeling results showed that the vapor maldistribution (which was not much, 52.8% to A panel and 47.2% to the B panel) on Tray 4 generated a large liquid maldistribution (67.3% to side A, 32.7% to side B) on tray 5 and 6, but by the time one reached trays 7 and 8 the pattern reverses itself, and the liquid maldistribution changes sides and recedes in magnitude.

Budget and schedule constraints precluded applying a similar analysis in the current case, but the results of the previous analysis (4) can be extended to it. In both cases we have a vapor-dominated distribution. There is one significant difference. In the wash section of the current tower (beneath tray 10) the liquid flow rates are much lower, so the resulting liquid maldistribution would be much more severe.

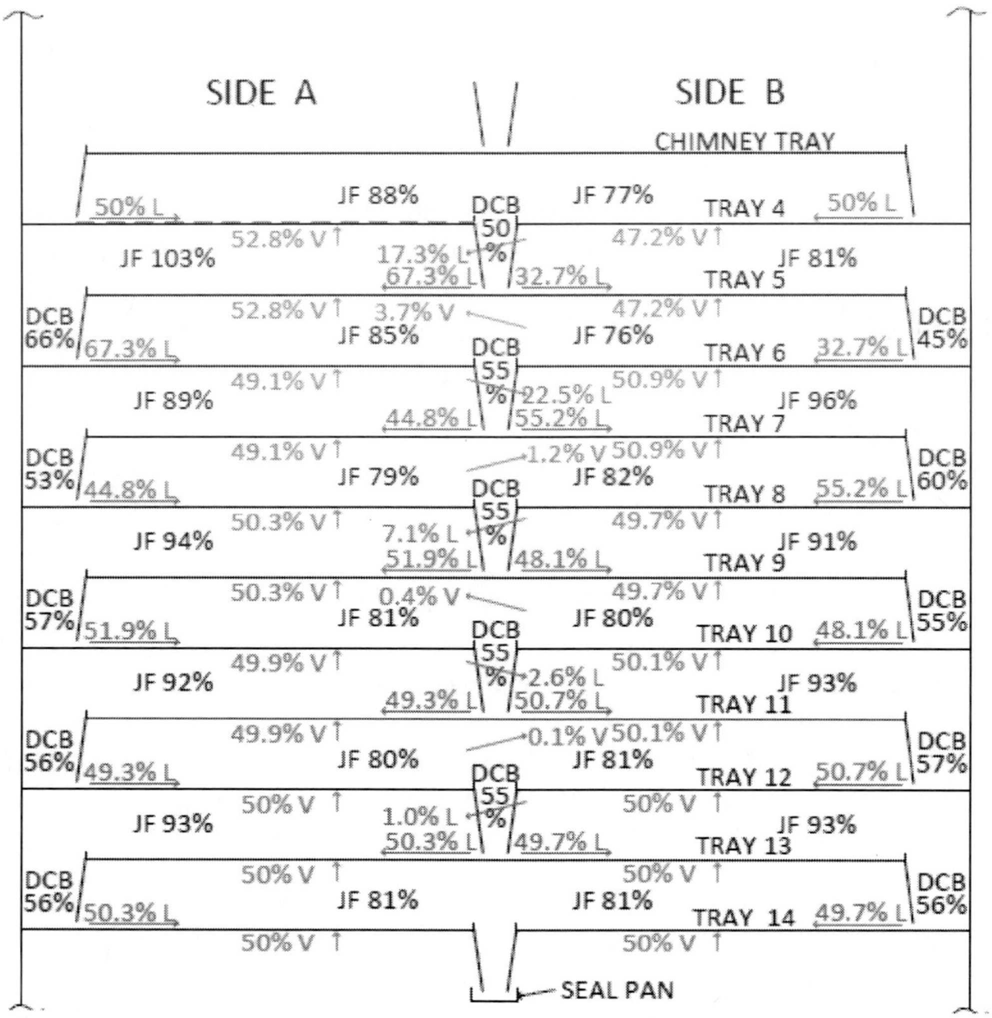

Figure 5. Application of the Fluor Maldistribution Model to analyze the distribution pattern beneath Tray 4 in another crude tower for a scenario in which 90% of the valve floats on Tray 4 were missing on side A, and all floats were present on side B. Legend: V = Vapor flow rate (mass units), L = Liquid flow rate (mass units), JF = %Jet Flood, DCB = Downcomer backup

NEW THEORY: PLUGGING ON ONE OR MORE TRAYS

This theory postulates that one or more trays were plugged. Plugging on tray 10 will explain the flooding above tray 10 seen on the scans. However, to be valid it also needed to explain the maldistribution, the unpredictable nature of the flood, and the vapor and liquid sensitivities of the flood.

Referring to Figure 4, complete pressure equalization between sides A and B can be counted on above tray 10 and above tray 6, i.e., $P10_A = P10_B$, and $P6_A = P6_B$, where P signifies pressures, psia, the numbers 6 and 10 signify the tray number, and the subscripts A and B signify the side per Figure 4.

Normally, one would expect vapor equalization also on tray 8. However, on that tray the sump and the outlet weir pickets block most of the equalization area. We estimate that the vapor equalization window above tray 8, including the gap between the picket fences and the sump, and including the area between the pickets, to be a total of about 3 ft^2. This is less than half the hole area on the tray (about 7.7 ft^2). So only a partial equalization is likely to occur above tray 8, and $P8_A$ can be lower or higher than $P8_B$.

Now let us postulate that tray 10 panels on side B are plugged to a greater extent than on side A. The pressure drop on side B will be higher, which will divert more vapor to side A until the pressure drop on trays 6-10 equalizes between sides A and B (i.e., $P6_A$ - $P10_A$ = $P6_B$ - $P10_B$). Assuming trays 6-9 are not plugged, the pressure drop on side B of trays 6-9 will be less than on side A because of their lower vapor loads. To keep the total pressure drops the same on the A and B sides of trays 6-10, $P10_B$ will need to be larger than $P10_A$.

The downcomer backup equals the pressure drop of the tray above, plus the liquid head at the inlet of the tray below, plus the friction at the downcomer apron (5). This means that if the pressure drop is higher on panel $P10_B$, more liquid needs to flow into panel 9A to satisfy the downcomer backup equation in the center downcomer from tray 10.

This is analogous to the situation on Tray 4 in Figure 5. Due to the popping of the valve floats on the A side, the B side had larger pressure drop, so the downcomer backup equation directed much more liquid, 67.3%, to flow to the A side, to balance the pressure drop difference. As stated above, due to the low liquid load in the current case, larger changes in liquid loads will be required to match the same change in vapor load. Consequently, an even a larger percentage of the liquid will be diverted to the A side, causing the B side to get close to drying up.

A similar mechanism can be postulated for plugging of one or more of the other trays 6-10, or for damage on one of these trays. The only premise is that the plugging is more severe on one side of that tray than the other. Due to the low liquid loads, the downcomer backup equation will direct large changes in the liquid split to counter even a relatively small vapor imbalance. What will vary, depending on the plugged or damaged tray(s), is the direction and magnitude to which the liquid will be preferentially directed.

The scenario may be further complicated by possible downcomer seal losses on the trays that approach dryness partially. The area under the center downcomers is 57% of the hole area. If it can find the path, the vapor has incentive to attempt rising up the downcomers. The inlet weirs ability to protect against seal loss depends on how much liquid comes into the inlet weir sumps compared to how much leaks out. When a tray gets close to drying, little liquid will come into the inlet weir sumps, and the downcomer seal is likely to be lost. This shifts the vapor distribution, and in turn the liquid distribution. Further, our sealing calculation shows that once the seal is broken the velocity of vapor in the downcomer is likely to exceed the system limit, preventing liquid descent. The liquid will pile up above the downcomer, and when enough builds up, eventually will dump. This will explain the erratic nature of the flood experienced in the tower.

Plugging of the active area parabolically raises the dry pressure drop, almost always resulting in vapor-sensitive downcomer backup flood. The liquid sensitivity came from tray 10. As this tray handles the pumparound liquid, as well as the diesel product flow, it has severalfold higher liquid loads than the low liquid load trays below. The liquid load affects the pressure drop of tray 10, and therefore, the flood.

In summary, the plugging theory was in line with the maldistribution observed in the gamma scans as well as the erratic nature of the flood and the vapor and liquid sensitivity of the flood. It became the leading theory.

TOWER INSPECTION

Based on the plugging theory, the tower was shut down for cleaning when the opportunity came.

- Trays 9 & 10 were found heavily fouled on the active panels. One could not see the holes. There was approximately 1/8" of tightly adhered sandy scale on all active surfaces of Tray 10 that could easily be removed with a metal scraper. On the underside of Tray 10 it was closer to 3/16" or ¼" thick. The top of Tray 9 was absolutely caked with foulant approximately ½" thick. It got worse as one neared the tower wall. The tower wall had no fouling and was clean. The downcomer box surfaces had some foulant, but the downcomers were fairly clean.
- The diesel draw on tray 9 was clear from mechanical obstruction.
- There was relatively less fouling on trays 8 and 7 compared to tray 9 and 10.
- Tray 6 had a light layer of debris/foulant on the panels and much cleaner than 10 and 9.
- No apparent physical or mechanical damage was noted for trays 6-10.
- Trays 11 and 12 were found spotless and clean (below the pumparound return).
- Trays 11 and up were clean and no sign of mechanical damage or heavy fouling.

a

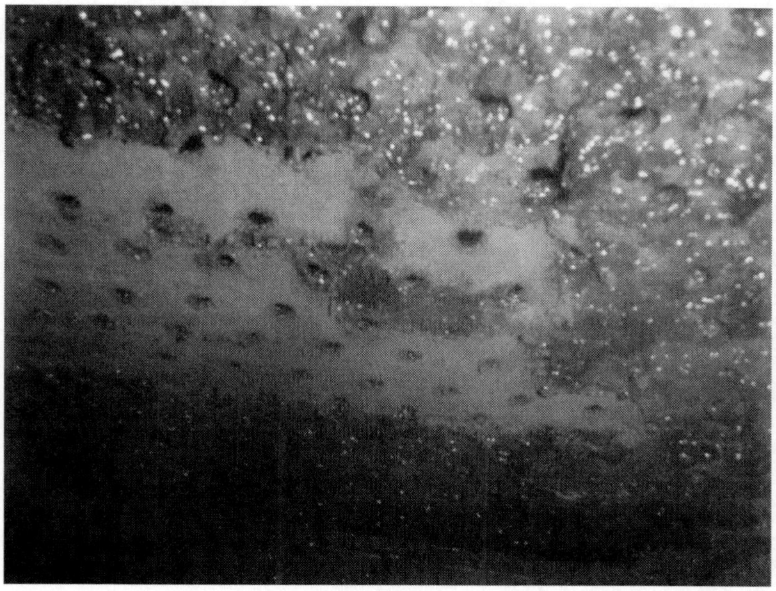
b

c

d

e

f

Figure 6. Tray photographs as seen when tower was opened in July 2022.

(a) Tray 10 top
(b) Tray 10 underside
(c) Tray 9 top
(d) Tray 8 top
(e) Tray 7 top
(f) Tray 6 top

NATURE OF THE FOULANTS

Lab analysis of the foulant showed ~8 wt% phosphorous, as well as excessive amount of FeS (~70-80%) in the deposit. It is important to note chemical cleaning (RTI) was used before opening the tower, which may have washed off some of phosphorous.

No analysis test was conducted for drilling mud because the X-ray deflection analysis showed small amounts of some parameters common for drilling mud but there was no bentonite, barite or hematite so this was not indicative of the presence of a water-based drilling mud.

Most experiences with phosphorous fouling have been in the jet pumparound or near the jet draw, but there have been some phosphorous-plugging experiences also in the jet-diesel section. The phosphorous compounds usually deposit at colder temperatures than those at the diesel draw, typically those near the jet draw. Also, deposits from sections fouled by phosphorous usually contain 10-50% phosphorous. The 8% phosphorous in the deposits of trays 9-10, as well as the observation that trays 11 and up, including the trays in the jet section, even trays 11 and 12 in the diesel pumparound, were clean, argue against phosphorous fouling.

The formation of FeS in a hot flowing oil is believed to involve two successive processes: the corrosion of iron by sulfur compounds and the decomposition of iron salts, for example, of naphthenic acids (8). The corrosion of iron by organic acids occurs on the surfaces of equipment. High temperatures favor FeS formation reaction, with temperatures such as that at the diesel draw (around 580°F) being well within the range. Most important, in the previous turnaround, trays 8-10 in the tower were replaced by in-kind, but the tray material was changed from 410 SS to carbon steel, which is far more prone to corrosion and FeS formation.

So, while the source of the plugging could be either FeS or phosphorous, or both, it is likely that FeS was a major actor.

TAKEAWAY

There are a number of invaluable lessons taught by the troubleshooting investigation:

- Flooded and non-flooded gamma scans of both active areas and downcomers are invaluable for correctly diagnosing tower problems.
- Quantitative analysis of gamma scans provides concise information on liquid maldistribution between panels of 2-pass or multi-pass trays.
- Factors like plugging or damage that cause a seemingly fairly small maldistribution of vapor between panels can lead to large variations in liquid splits to the panels.
- Theories should be tested against plant and gamma scans data to obtain correct diagnostics.
- Where fouling is expected, plugging-resistant tray designs should be applied. In wash sections, it is also important to ensure the tray design resists downcomer unsealing.

REFERENCES

1. Kister, H. Z., "Use Quantitative Gamma Scans to Troubleshoot Maldistribution on Trays", Chem. Eng. Progr., February 2013.
2. Harrison, M. E., "Gamma Scan Evaluation for Distillation Column Debottlenecking", Chem. Eng. Prog. 86 (3), p.37-44, March 1990.
3. Kister, H. Z., "Is the Hydraulic Gradient on Sieve and Valve trays Negligible?" Paper presented at the Topical Conference on Distillation, AIChE Meeting, Houston, Texas, 2012.
4. Kister, H. Z., and M. Olsson, "An Investigation of Premature Flooding in a Distillation Column", Chem. Eng., p.29, January 2019.
5. Kister, H. Z., R. W. Dionne, W. J. Stupin, and M. Olsson," Preventing Maldistribution in Multi-Pass Trays", Chem. Eng. Prog., p.32, April 2010.
6. Olsson, M., and H. Z. Kister, "Can We Count on Good Turndown in Two-Pass Moving Valve Trays", Chem. Eng. Prog., p.43, November 2018.
7. Kister, H. Z., P. M. Mathias, D. E. Steinmeyer, W. R. Penney, V. S. Monical, and J. R. Fair, "Equipment for Distillation, Gas Absorption, Phase Dispersion and Phase Separation", in D. W. Green and R. H. Perry "Perry's Chemical Engineers' Handbook" 9th Ed., Section 14, McGraw-Hill, New York, 2018.

8. Wang, W., and A. P. Watkinson, "Iron Sulphide and Coke Fouling from Sour Oils: Review and Initial Parameters", Proceedings of International Conference on Heat Exchanger Fouling and Cleaning 2011 Crete Island, Greece, June 5-10, 2011.

The absolute necessity of getting them right:
Surface tension predictions for high-pressure column operations

Ruan de Villiers [a], Attilio Praderio [b], Hendrik A. Kooijman [a], Ross Taylor* [a]

[a] Department of Chemical and Biomolecular Engineering
Clarkson University
Potsdam, New York 13699-5705, USA

[b] Consultant
8627 Crescent Valley Lane, Humble, TX 77346

* *Corresponding author's email:* rtaylor@clarkson.edu

Presented at the Kister Distillation Symposium 2023
AIChE Spring Meeting, Houston, Texas, March 2023

Introduction and Motivation

> *"The importance of surface tension in design of separation equipment can only be understated. Over prediction of surface tension, particularly when significantly below 10 dyn/cm, can lead to significant undersizing of vapor–liquid separations equipment.[169]"*

The quote with which we begin this introduction comes from a recent paper by Gupta et al. (2023). The number 169 that appears as a superscript at the end of that quote is a reference to an article by McCarley (2021).

This led us to ask this question (the first of many):

1. How low can surface tension go?

Surface tension (denoted often in what follows by σ) is unlike most other properties in that it is a property that requires the existence of two different phases (density, viscosity, and thermal conductivity are single phase properties). Surface tension decreases as pressure and temperature increase, reaching zero at the critical point (beyond which only a single phase exists). Thus, low values of surface tension are more likely to occur in distillation operations carried out at high pressure. With that in mind we modeled a number of processes that include high pressure distillation columns; Table 1 includes a summary of the processes that we considered. We note, in particular, the last two rows of Table 1 are for an unusually high pressure C_4 splitter. These pressures are much higher than the operating pressure of a typical C_4 splitter; they do, however, represent pressures at which experimental tests are carried out with this particular mixture.

Table 1: Selected High Pressure Column Operations

Process	Key components	Pressure	Surface tension (dyne/cm)
Demethanizer from natural gas train	$CH_4 - C_2H_6$	25 atm	2 – 9
CO_2 – Ethane separation	$CO_2 - C_2H_6$	23.7 atm	2 – 8 ; 2 – 3
Ortloff process	$CH_4 - CO_2$	4.3 MPa	3 – 10
Nitrogen rejection unit	$N_2 - CH_4$	2650 kPa	0.5 – 2
C_4 splitter	$i\text{-}C_4H_{10} - n\text{-}C_4H_{10}$	658 kPa	10
C_4 splitter	$i\text{-}C_4H_{10} - n\text{-}C_4H_{10}$	400 psia*	1
C_4 splitter	$i\text{-}C_4H_{10} - n\text{-}C_4H_{10}$	500 psia*	0.1 – 0.3
			*Test pressure

It is clear from the result in Table 1 that the surface tension may sometimes be significantly below 10 dyne/cm. (We have, however, encountered cases where the surface tension is less than 0.1 dyne/cm.)

This brings us to our next questions:

2. From where did we get those values of σ?
3. Are they realistic (most especially the very low values)?
4. Where and how does σ impact distillation design?
5. What are the consequences of incorrect estimates of σ?

We continue with our answers to these questions.

2. From where did we get those values of σ?

The answer to this question is that they came from a model; more precisely, a model for *mixture* surface tension. In Table 2 we list the availability of surface tension models in some process simulation tools. We note that ChemSep is not a process flowsheet simulation tool as are the others in this table. ChemSep has two essential functions; it models columns and can provide physical and thermodynamic properties to a CAPE-OPEN compliant software system. Note that the availability or non-availability is not an indication of superiority. What matters is the quality of the available models and that we shall address later in this paper.

Table 2: Surface tension models available in some process popular simulation tools

Mixture Surface Tension Model	ChemSep+	Aspen*	ChemCAD*	Symmetry*
Molar average	✓	✓		✓
Power law	✓	✓		✓
Weinaug – Katz	✓	✓	✓	
Lee–Chien	✓			
Winterfeld – Scriven – Davis	✓			
Brock – Bird	✓			
Other general models	✓			✓
Other special models		✓		✓

3. Are our estimates of σ realistic?

Models for predicting the surface tension of a given system can be divided into two broad categories: pure-component (PC) models and mixture models. As will be demonstrated in what follows, the strategies for estimating PC surface tensions cannot always be extended to a multicomponent form, and when they can, those extensions do not necessarily permit the model to be used for any and all mixtures at any or all combinations of temperature, pressure, and composition.

Estimating the surface tension of a pure fluid is a relatively straightforward procedure; data for most species can be adequately fitted with a simple function of temperature such as:

$$\sigma = A + \exp\left(\frac{B}{T} + C + DT + ET^2\right) \quad (1)$$

This is the functional form adopted in ChemSep (as well as in other systems). Figure 1 shows fits to data for a small selection of chemicals that appear in many high pressure separations.

It is important that any model or correlation of surface tension exhibit the correct limiting behavior: as T approaches T_c. that is, σ approaches zero. That this is the case for the function shown above is evident in Figure 1.

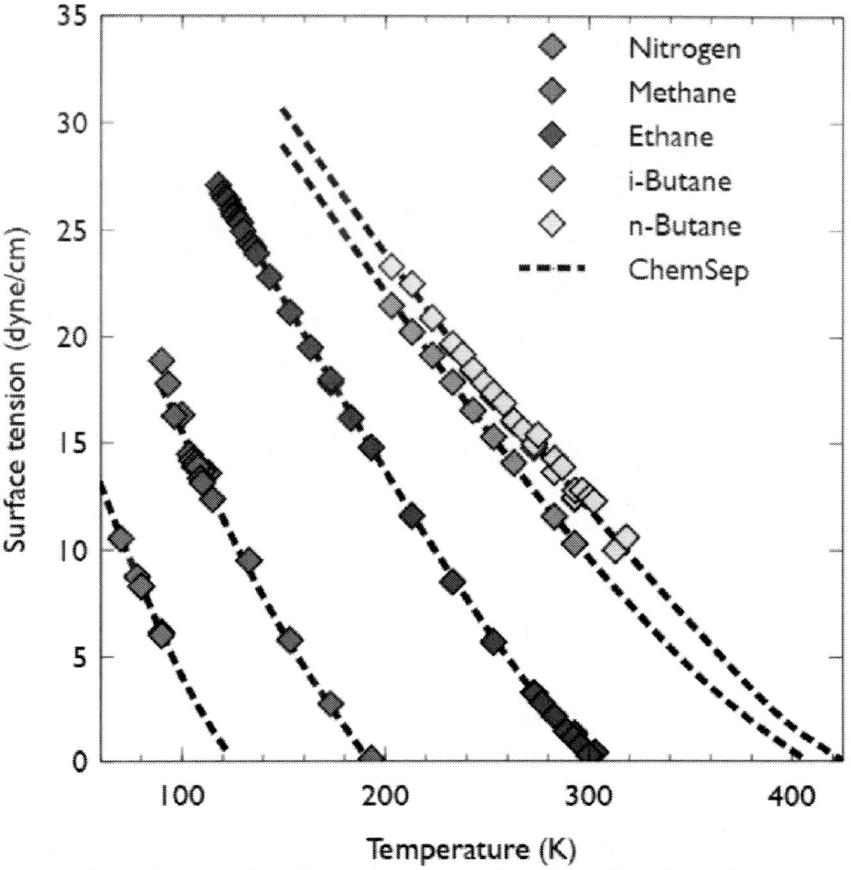

Figure 1: Correlation of surface tension used in ChemSep for a few pure components.

Simple correlations such as that shown in Figure 1 can provide good fits to experimental data, but are lacking in predictive quality if no data is available. In this case, one might choose to use the predictive method of Brock & Bird (1955), which uses only the critical properties and boiling point of a species to estimate surface tension:

$$\sigma = 4.601 \times 10^{-7} \times (P_c)^{2/3}(T_c)^{1/3}Q(1-T_r)^{11/9} \quad \textbf{(2)}$$

$$Q = 0.1207\left(1 + \left(\frac{T_{b,r}}{1-T_{b,r}}\right)\ln\left(\frac{P_c}{101325}\right)\right) - 0.281 \quad \textbf{(3)}$$

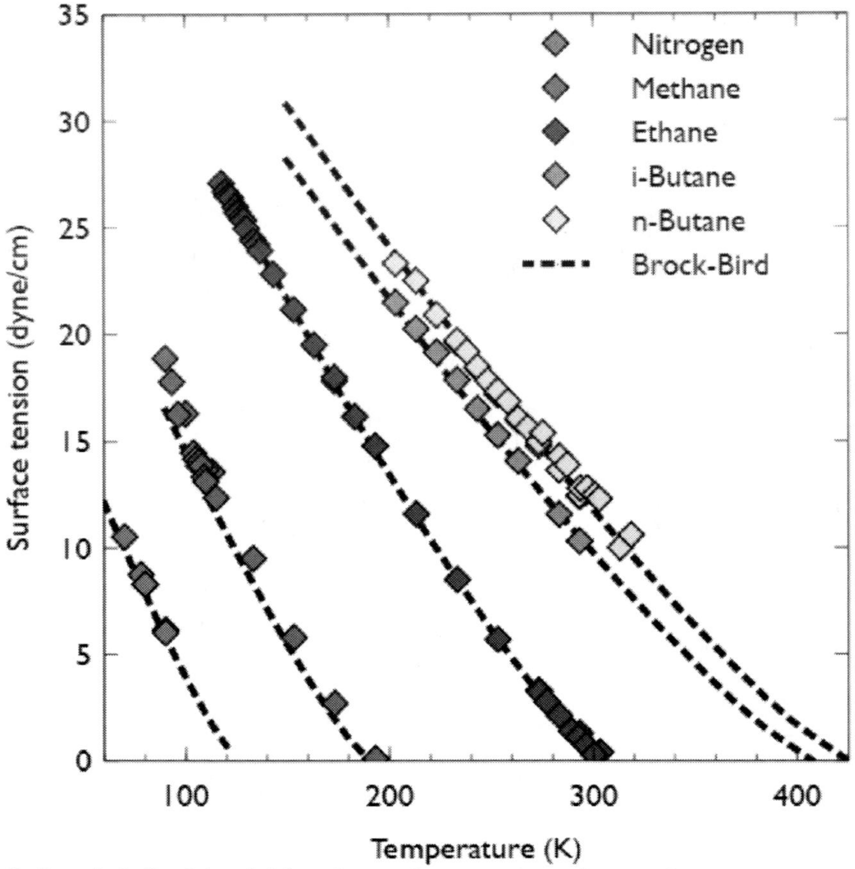

Figure 2: Brock & Bird model for the surface tension of several pure components.

The model is demonstrated in Fig. 2; for the chemicals selected here the Brock – Bird model is only marginally less accurate than was the correlation that was fit to data! This model can also be used for mixtures by using pseudo-critical mixture properties in place of the pure component values. (It would be wiser to use true critical properties but those are not likely easy to obtain.)

The simplest method for estimating the surface tension of a mixture is to use a mole fraction weighted average of the pure component values:

$$\sigma = \sum_{i=1}^{c} x_i \sigma_i \tag{4}$$

There are times when such a simple model is entirely appropriate and perfectly adequate, but it is not often a good model for surface tension as we can see in Fig. 3. The method is highly inaccurate when conditions exceed the critical point of one (or more) of the individual components (as is the case in Fig. 3). When that happens, the model is averaging something (the surface tension of propane) with nothing (the surface tension of methane). Clearly this is not an effective approach in these circumstances. However, it can happen that this is the only model available in some software systems; this despite it being the worst, on average, possible model.

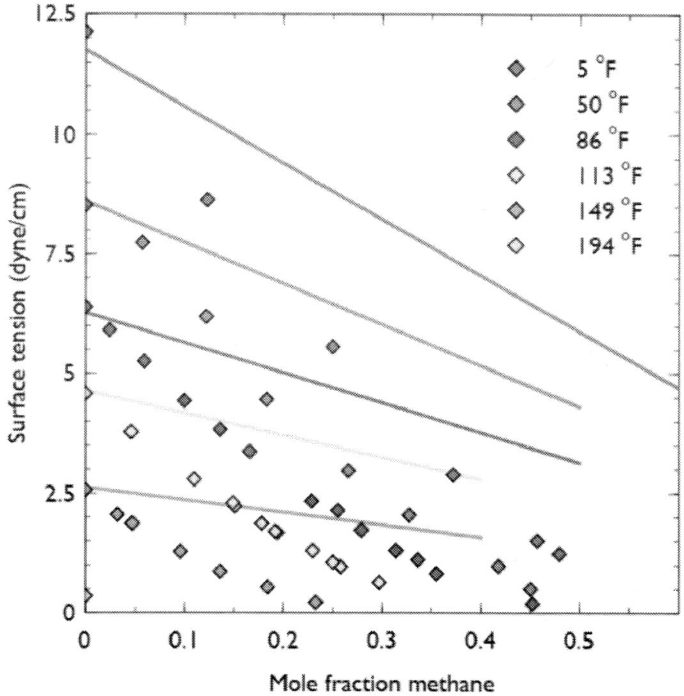

Figure 3: Molar average mixture surface tension for methane – propane.

We can improve the simple molar average with the almost equally simple power law model

$$\sigma^r = \sum_{i=1}^{c} x_i \sigma_i^r \tag{5}$$

In which r is an exponent whose value can best be determined by comparing the model to data. Figure 4 shows the surface tension of the same methane – propane mixture that we saw in Fig. 3, here using Equation (5) and with $r = 0.3$. Also shown in Fig. 4 is the methane – ethane system with $r = 0.65$ (that value demonstrated by D. Summers (2023) to provide a good fit to the data at 173 K). It is readily apparent that the model is an improvement over the simple mole fraction average (equivalent to $r = 1$) but it is not much of an improvement!

Further improvements can be obtained by making r a function of temperature as we can see in Fig. 5. The utility of this model is quite limited because the exponent is not knowable in advance and depends on temperature in ways that are impossible to predict.

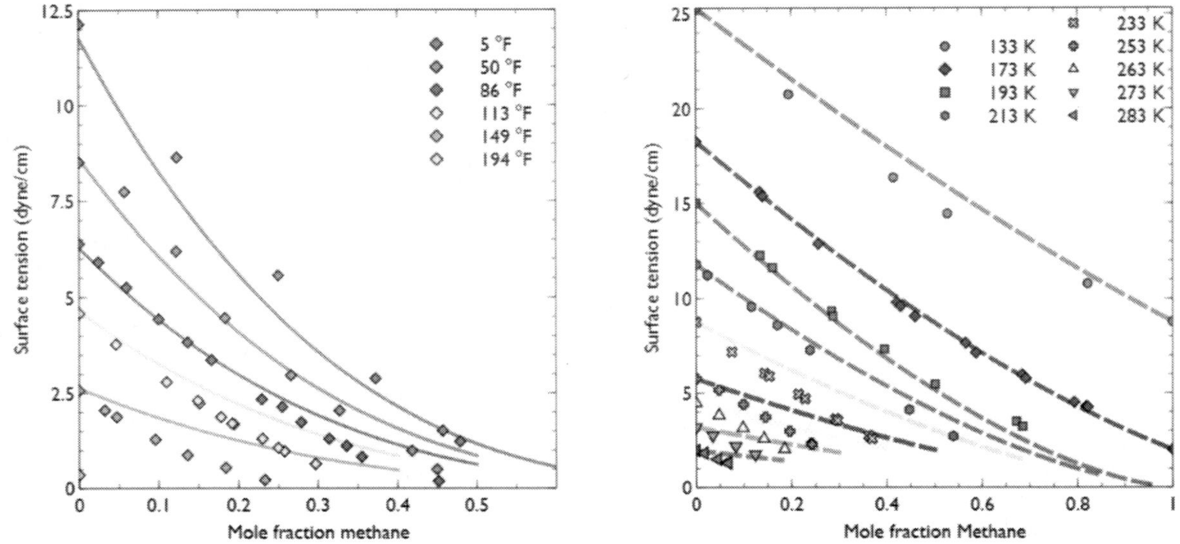

Figure 4: Power law mixture surface tension for methane – propane (left) with r = 0.3 and for methane – ethane (right) with r = 0.65

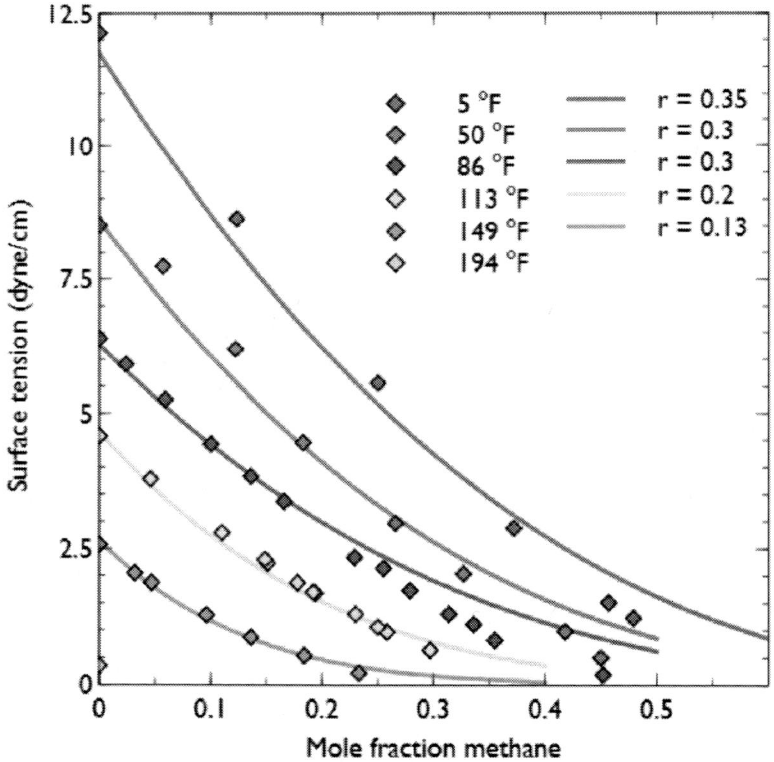

Figure 5: The same methane – propane data as Fig. 4 (left) with the power law exponent now allowed to vary with temperature.

The model of Winterfeld et al. (1978) requires pure component surface tensions and pure component densities:

$$\sigma = \sum_{i=1}^{c}\sum_{j=1}^{c}(\rho^*)^2 \frac{x_i}{\rho_i^L}\frac{x_j}{\rho_j^L}\sqrt{\sigma_i\sigma_j} \tag{6}$$

$$\frac{1}{\rho^*} = \sum_{i=1}^{c}\frac{x_i}{\rho_i^L} \tag{7}$$

The model is compared to data for methane – propane in Fig. 6. However, as this figure shows, it is not universally useful. The model is not suitable for mixtures that contain compounds above their own critical points.

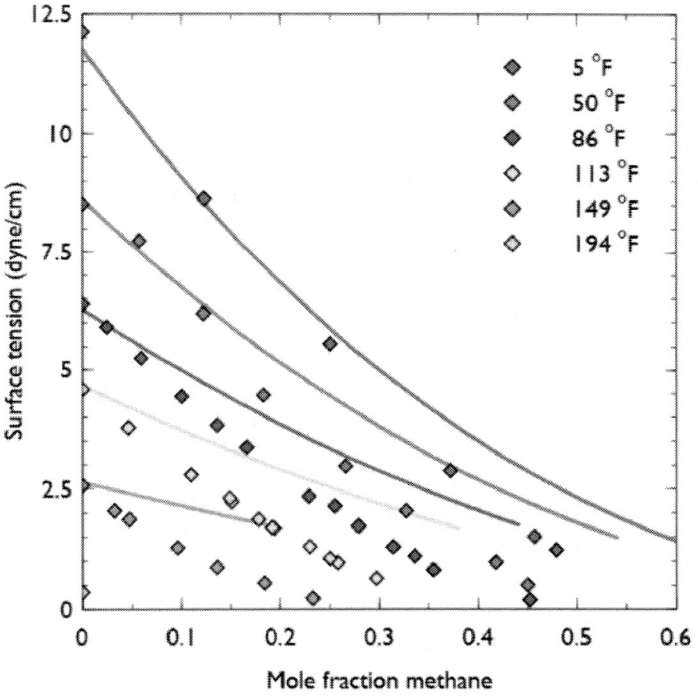

Figure 6: Winterfeld – Scriven – Davis model for methane – propane.

In 1923 D.B. Macleod presented a simple relationship between surface tension and density:

$$\sigma = C(\rho_t^L - \rho_t^V)^4 \tag{8}$$

In which C is a parameter. Sugden (1924) rearranged this equation to:

$$\sigma^{1/4} = \wp^L\rho_t^L - \wp^V\rho_t^V \tag{9}$$

and named the quantity denoted by $\wp$ the parachor. The Merriam-Webster dictionary defines parachor as "*an empirical constant for a liquid that relates the surface tension to the molecular volume.*" The parachor has some justification in statistical mechanics (Boudh-Hir & Mansoori, 1990) and evidence suggests that $\wp$ is not a strong function of temperature.

The Macleod – Sugden model is quite good for many pure components; evidence appears in Fig. 7. The parachor can be estimated from available surface tension data or predicted from group contribution models. Also, as required, the model has the correct limiting behavior approaching the critical point ($\sigma \longrightarrow 0$). One potential drawback is that it needs accurate density data which may not be available, especially near critical points. (We will demonstrate this point below.)

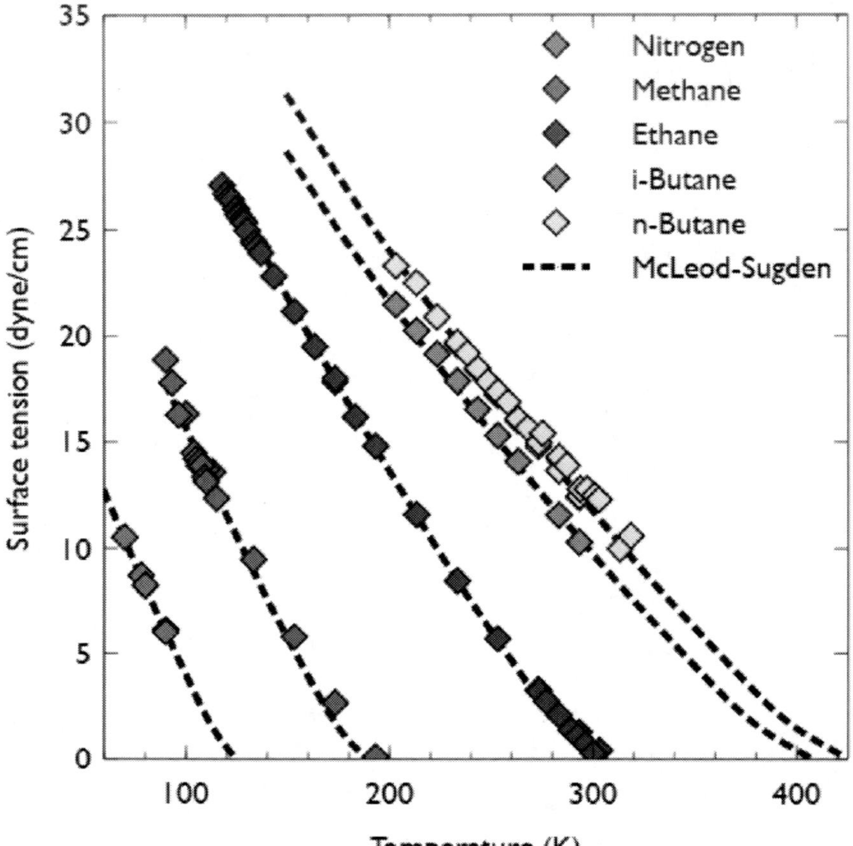

Figure 7: Macleod – Sugden model for the surface tension of several pure components.

The Macleod – Sugden model was adapted for mixtures by Weinaug & Katz (1943); Equation (9) remains in use but the parachors are estimated from:

$$\wp^L = \sum_{i=1}^{c} \wp_i^L x_i \tag{10}$$

$$\wp^V = \sum_{i=1}^{c} \wp_i^V y_i \tag{11}$$

Other parachor-based models are adapted from that of Weinaug & Katz. The Lee – Chien (1984) model for example, uses pseudo-critical parameters that are mole fraction averages of the corresponding pure component properties

$$\wp^* = \frac{(A_c^*)^{1/4} V_c^*}{B^*} \quad \textbf{(12)}$$

$$A_c^* = 4.601 \times 10^{-7} \times (P_c^*)^{2/3} (T_c^*)^{1/3} \times (0.133\alpha_c^* - 0.281) \quad \textbf{(13)}$$

B^* is specified for a few important chemicals, otherwise estimated from the critical compressibility.

Comparisons of these two models for methane – propane are shown in Fig. 8. Neither model is "always better" than the other. Note, however, how the Lee – Chien model (on the right in the figure) shows poor predictions of the pure propane surface tension.

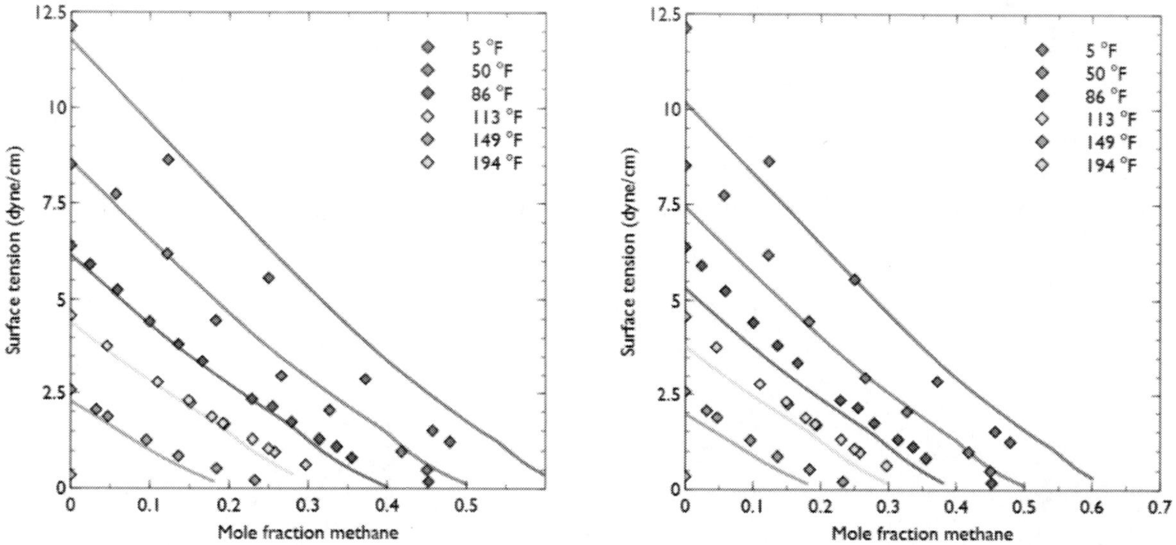

Figure 8: Two parachor-based models for surface tension. Data is for the binary mixture methane – propane. Left is the Weinaug – Katz model; right is Lee – Chien.

We can summarize the strengths and weaknesses of parachor-based models as follows:

- Do NOT directly use pure component densities or σ_i
- Possess the correct behavior when approaching critical point ($\sigma \longrightarrow 0$)
- Can be used when some components are above their own critical points
- Model somewhat sensitive to liquid density estimation method
- Needs accurate liquid density near (pseudo)-critical point

On that last point we must note that accurate density estimates are not always available. We can see this in Fig. 9 which shows the molar volume ($1/\rho$) for two ethane – *n*-decane mixtures. Figure 10 show how the method selected to estimate the liquid density impacts surface tension from the Weinaug – Katz model.

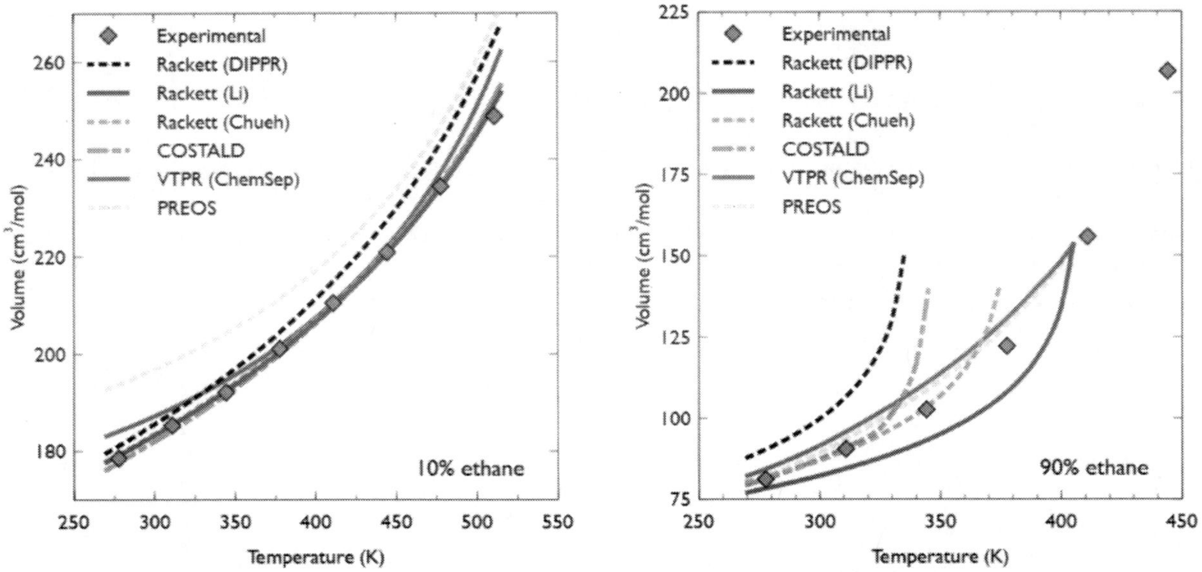

Figure 9: Molar volume (1/ρ) for ethane – *n*-decane at two different mixture compositions.

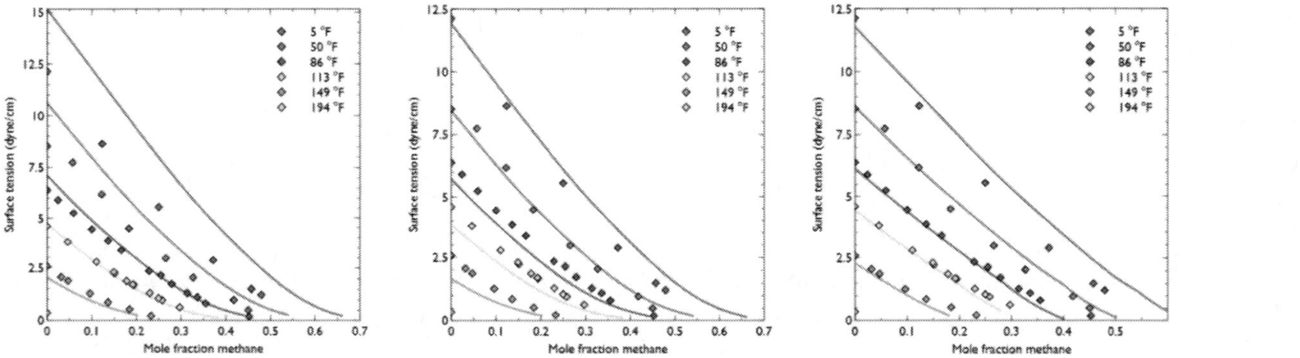

Figure 10: Weinaug – Katz model for methane – propane using standard (left) & volume-translated (center) Peng-Robinson EOS for liquid density. Right plot uses the COSTALD model for density.

Figure 11 shows the predictions of the models considered so far in this paper for a mixture of five hydrocarbons. None of the models is really good enough here, except, occasionally by accident!

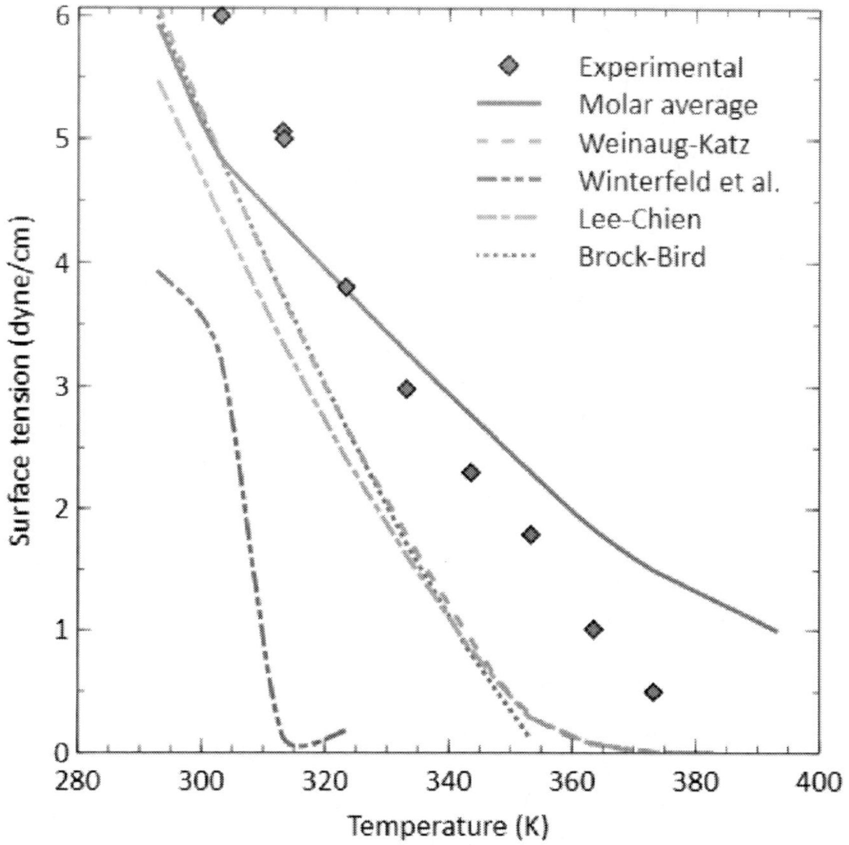

Figure 11: Surface tension for a mixture C_1, C_2, C_3, C_4, C_7 using a variety of mixture models. Data from Horstmann et al. (2010).

We may summarize our findings thus far as follows:

- No model in common use performs well all the time!
- Many models should not be used when the system temperature exceeds the critical temperature of a species in the mixture!
- Many simulators have only a limited choice of models, some of which are not very good much of the time.
- Some simulators do not readily disclose what mixture σ model they use!
- Or, if they do tell you, it is well hidden.
- Often the default choice is the molar average, often, but not always the worst possible choice.

4. Where and how does σ impact distillation design?

Surface tension appears in a very large number of sizing and design correlations for both trayed and packed columns. Below we present the equations for a small sample of representative calculations. Surface tension is highlighted in red.

$$u_{N,flood} = C_{SB}\left(\frac{\sigma}{20}\right)^{0.2}\left(\frac{\rho_L - \rho_V}{\rho_V}\right)^{1/2} \tag{14}$$

$$C_1 = \left(\frac{1.4}{1 + 1.4\sqrt{\Delta\rho/\rho_V}}\right)\left(\frac{4g}{C_D}\right)^{1/4}\left(\frac{\sigma\Delta\rho}{\rho_V^2}\right)^{1/4} - 1.4L_s \tag{15}$$

$$C_{s,ult} = min(C_1, C_2) \tag{16}$$

$$C_2 = \left(\frac{1.12}{1 + 1.4\sqrt{\Delta\rho/\rho_V}}\right)\left(\frac{4g}{C_D}\right)^{1/4}\left(\frac{\sigma\Delta\rho}{\rho_V^2}\right)^{1/4} \tag{17}$$

$$\frac{a_e}{a_p} = 0.498\left(\frac{\sigma^{0.5}}{Z^{0.4}}\right)\left(\frac{\mu_L L}{\rho_L \sigma g_c}\right)^{0.392}\left(\frac{6G}{a_p \mu_G}\right)^{0.392} \tag{18}$$

$$\frac{a_e}{a_p} = 1.5\left(a_p d_h\right)^{-0.5}\left(\frac{u_L d_h}{v_L}\right)^{-0.2}\left(\frac{u_L^2 \rho_L d_h}{\sigma}\right)^{0.75}\left(\frac{u_L^2}{g d_h}\right)^{-0.45} \tag{19}$$

$$\frac{Đ_{1,2}^{\circ}\mu_2}{T} = 8.93 * 10^{-8}\left(\frac{V_1^{1/6}}{V_2^{1/3}}\right)\left(\frac{\wp_2}{\wp_1}\right)^{0.6} \tag{20}$$

$$\wp = \sigma^{1/4}V \tag{21}$$

Equation (14) is Fair's (1982) superficial tray flooding velocity (see Kister, 1992), Equations (15 – 17) are the Stupin – Kister (2002) ultimate capacity correlation. Equation (18) is the Bravo – Fair (1982) correlation for effective interfacial area of random packing. Equation (19) is the Billet – Schultes (1993, 1999) interfacial area correlation for packing. Equations (20 – 21) are the Tyn – Calus (1975) correlation for the diffusion coefficients in a dilute binary mixture.

The models presented above (as well as many others) ultimately derive any success they may have from observed (measured) performance data. It is, therefore, important to know the range of any parameter that appears in these (or other) equations. By way of example, we summarize, in Table 3, the ranges for the physical properties that we used in the creation of the interfacial area correlations by Bravo & Fair (1982) and by Billet & Schultes (1999). It can be seen that the latter correlation covers a much wider range of property values than does the former. (This should not be taken as a criticism of the model of Bravo & Fair; that of Billet & Schultes was developed many years later and could, therefore, take advantage of significantly more data.)

Table 3: Physical properties were used in correlation development

Property	Bravo – Fair (1982)	Billet – Schultes (1999)
Gas density (kg/m³)	0.26 – 28.2	0.07 – 97
Liquid density (kg/m³)	497 – 833	361 – 1237
Gas viscosity ($10^6 \times m^2/s$)		0.14 – 126
Liquid viscosity ($10^6 \times m^2/s$)		0.14 – 1.66
Surface tension (dyn/cm)	**6* & 15 – 27**	**0.7 – 74**
* Data from Billet (1967) for *i*-Butane – *n*-Butane at 165 psia Details of source not provided by Bravo (1981)		

In his Master's thesis José Bravo (1981) identified the distillation test systems used in the development of his model. Note that the *i*-Butane – *n*-Butane pair is the only one at a moderately high pressure, but even this pressure is not high enough to lead to very low surface tensions.

Table 4: Systems used for development of Bravo-Fair correlation

Test Mixture	Pressure(s)
Cyclohexane – *n*-Heptane	5, 24 psia
n*-Butane – *i*-Butane	**165 psia**
i-Propanol – Water	14.7 psia
Ethylbenzene – Styrene	1, 7 psia
Methanol – Ethanol	14.7 psia
Ethanol – Water	14.7 psia
n-Heptane – Toluene	14.7 psia
* Surface tension data for this mixture not available in Dortmund Data Bank or NIST Thermophysical property collections	

5. What are the consequences of incorrect estimates of σ?

Table 5 summarizes some design calculations carried out for a demethanizer using two different sets of models. Note the very different diameters for these two designs! An inevitable consequence is that the Total Annualized Cost (TAC) of Design I is much less than the TAC of Design II. A more important concern, for us, is this: Will either design work? If we fix the diameter and flip the design models then Design I will not work and Design II will work (or so the simulations tell us). The surface tension is not totally to blame for these differences, but it is part of the story.

Table 5: Two different designs for some demethanizer column

Design model	Design I	Design II
Pressure drop and flood	Leva GPDC	Billet – Schultes
Design criterion	75% of flood	75% of flood
Mass transfer coefficient	Bravo – Fair	Billet – Schultes
Diameter	12 ft	16 ft

Figure 12 shows the diameter estimated from the Fair flooding correlation relative to that that would be calculated if the surface tension was 20 dyne/cm. Note that this says nothing about the validity of the underlying model. The Fair correlation is, however, very widely cited, especially in textbooks.

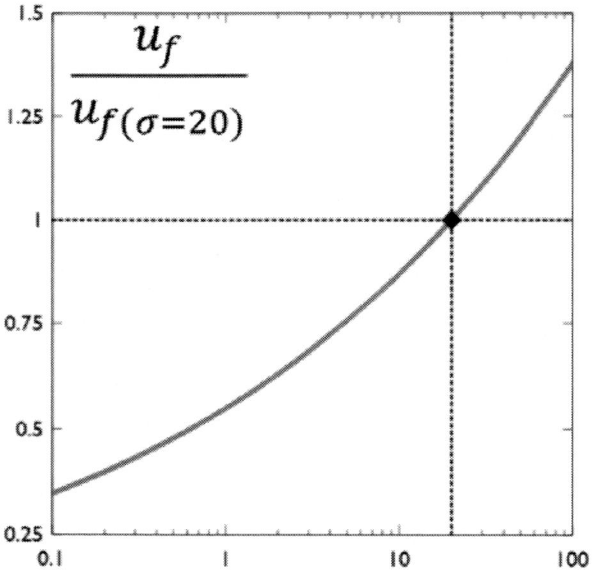

Figure 12: Relative flooding velocity from Fair model.

But this is not the correct comparison. One needs to compare results at the actual value of surface tension (not likely to be known precisely, of course) with the design that results from the estimated value of surface tension. Further, we need some idea of the uncertainty in the estimated surface tension.

That uncertainty is greater for smaller values of surface tension. Figure 13 (left) shows surface tension data for a binary mixture together with the predictions of two models, neither of which is very good. By looking at plots similar to the one shown in Fig. 13 (left), we observed that it is possible for models to overestimate the surface tension, sometimes by very significant amounts. The overestimate is shown in Fig. 13 (right). For example, Fig. 13 (right) assumes that if the surface tension is 0.1 dyne/cm it is possible to obtain an estimate of 1 dyne/cm (it is actually 4 dyne/cm). These are what we consider possible plausible overestimates; we do not mean that a model IS this bad, but that it MIGHT be this bad (it could also be worse). This is the basis for the calculations done later.

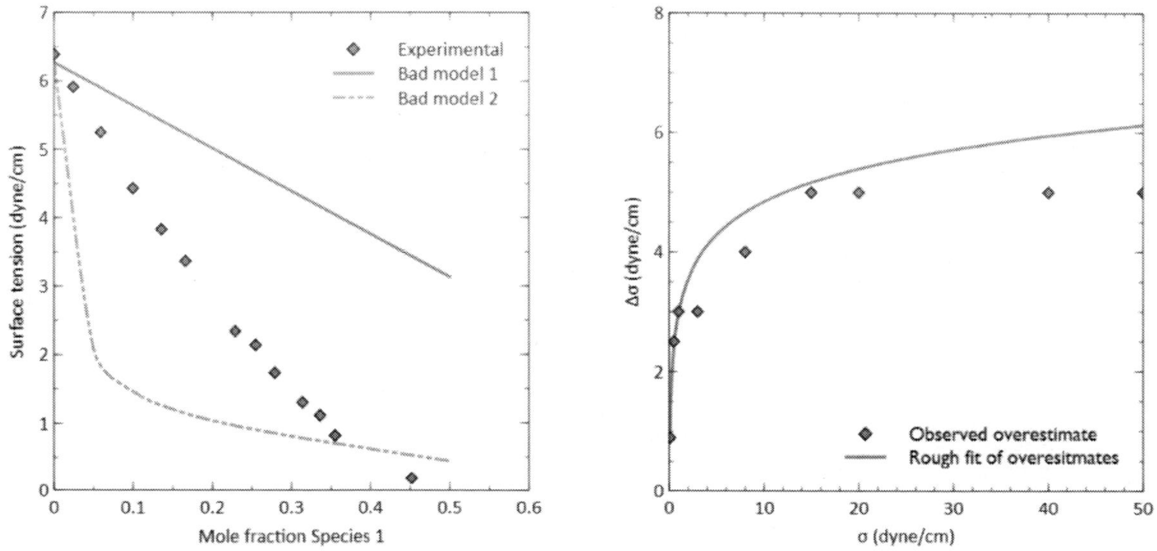

Figure 13: Two bad predictions of surface tension for a real binary (left) & plausible overestimate (right).

The plausible overestimate "data" can be very approximately fit by Equation (22):

$$\Delta\sigma = 3 - \frac{0.02}{\sigma} + 0.8\ ln(\sigma) \qquad \textbf{(22)}$$

Armed with this admittedly very approximate model for the possible error in the estimated surface tension we return to the calculation of the flooding velocity. Figure 14 shows the relative diameter calculated from Fair's tray flooding model. The purple line is the same curve shown in Fig. 13 (right). Note that the x-axis here is logarithmic rather than linear. The cyan colored curve is the relative diameter. That is, the diameter using the actual surface tension to the diameter at the presumed incorrect prediction. Thus, at a surface tension of 1 dyne/cm the diameters are about 15% different. We caution that this presumes the validity of the underlying model about which these calculations say nothing.

Figure 15 shows the shape of the interfacial area correlations from the Bravo – Fair and Billet – Schultes correlations. It is immediately evident that these two curves have quite different shapes, even though they are theoretically describing the same concept: the effective interfacial area of packing available for mass transfer.

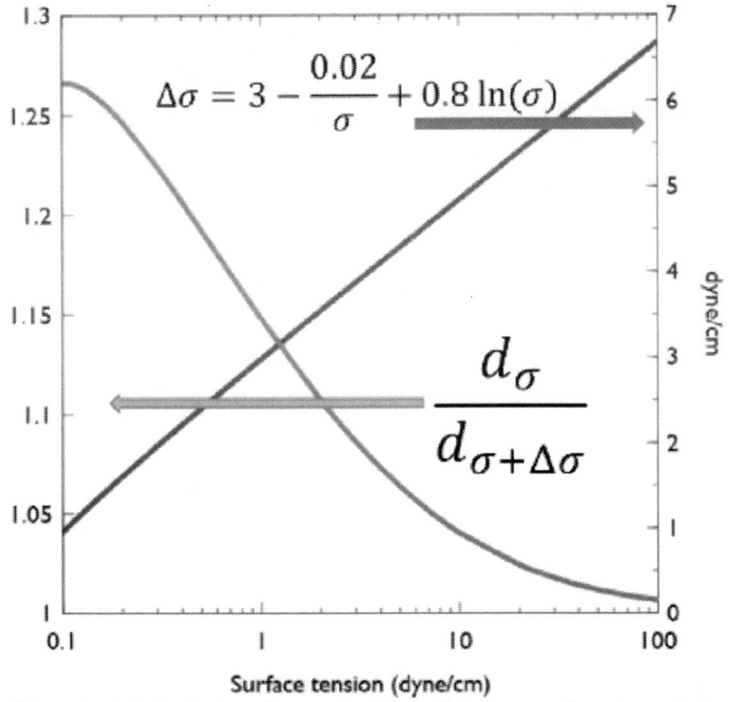

Figure 14: Relative diameter from Fair's tray flood model.

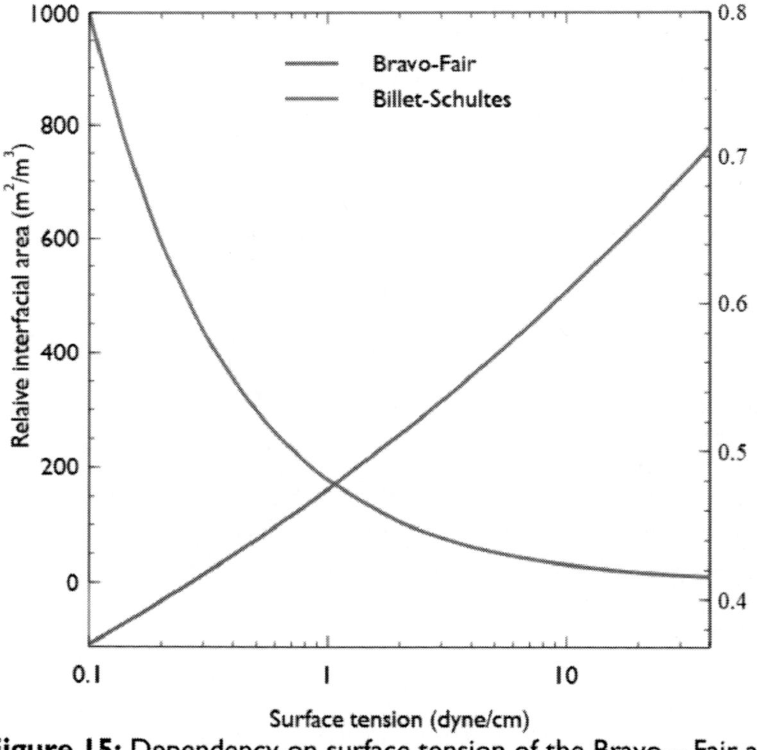

Figure 15: Dependency on surface tension of the Bravo – Fair and Billet – Schultes models for interfacial area in a packed column.

Figure 16 shows the actual interfacial area calculated from the Billet – Schultes model for a high pressure demethanizer. It is vital to note that Billet & Schultes impose an important restriction on their correlation: Thou shalt not use a value of σ less than 30 dyne/cm; if a surface tension of < 30 dyne/cm is predicted, then a value of 30 dyne/cm is used instead. This is what Billet and Schultes have to say:

> *"the extremely low surface tension of the liquid phase at high pressures leads to a large calculated interface for mass transfer. If, however, one takes into account the experimentally confirmed fact that, below a lower limit for the surface tension of 30 dyne/cm, the surface of metals, plastics or ceramic is completely wet, the influence of the surface tension on the wettability of dumped or arranged packings is lost. This was confirmed by the measurements at high pressures of up to 30 bar. The results of the Measurements agreed with the calculated results well if a minimum value of 30 dyne/cm is used for the surface tension."*

The broken line in Fig. 16 shows the Billet & Schultes model if one ignored their stipulation. It is evident that the interfacial area so computed is quite unrealistic, thus underscoring the importance of following their advice.

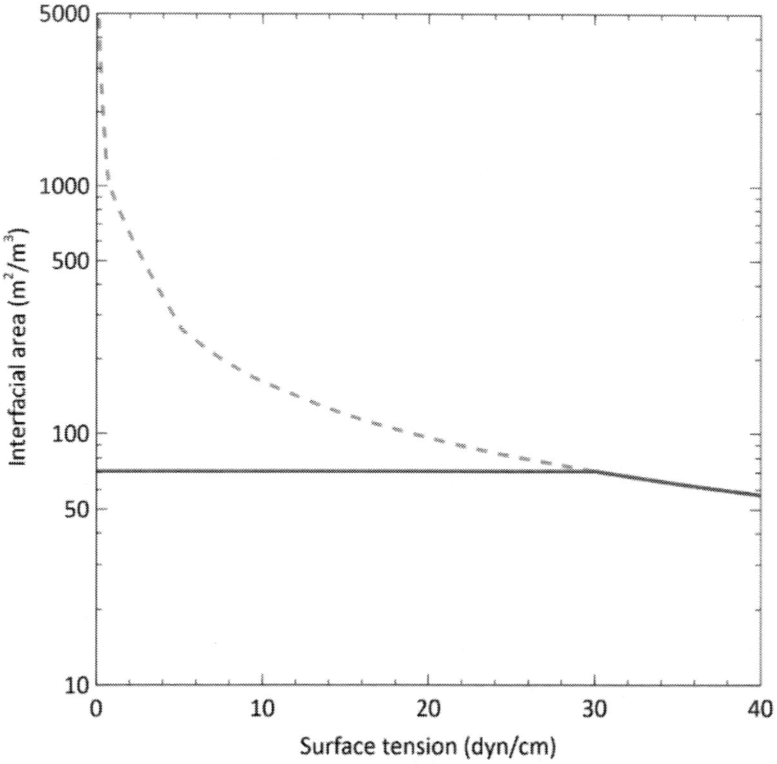

Figure 16: Plot of the Billet – Schultes correlation for a demethanizer. The dotted line uses the predicted values of surface tension, while the solid line uses the cutoff surface tension of 30 dyn/cm.

The contradictory shapes of Bravo – Fair and Billet – Schultes along with the internal surface tension limit imposed by Billet & Schultes may raise the concern that at least one, if not both, of the models are fundamentally incorrect at some level. Of course, "incorrect" or not, these models work well enough to continue to enjoy wide use among engineers who design real distillation columns that are actually constructed and operated successfully, so perhaps the issue is not with the models themselves but with how they are represented in Figure 15.

Underlying the creating Figs. 12 and 14–16 is an assumption that the surface tension can be varied independently of all other thermophysical properties. While it may suffice for the purposes of demonstrating the shape of Equations 5 and 6, it is necessarily inaccurate because surface tension is *not* independent of other fluid properties. In fact, it is strongly correlated to density, specifically the *density difference* between the vapor and liquid phases. Figure 17 shows the experimental surface tension data of Baidakov et al. (2012, 2013) against density *difference* for mixtures of two binary chemical pairs at various temperatures. (There is a similar but slightly weaker correlation with liquid density alone.) This relationship is described by an equation of state thus emphasizing the importance of thermodynamics in distillation; it isn't just for VLE.

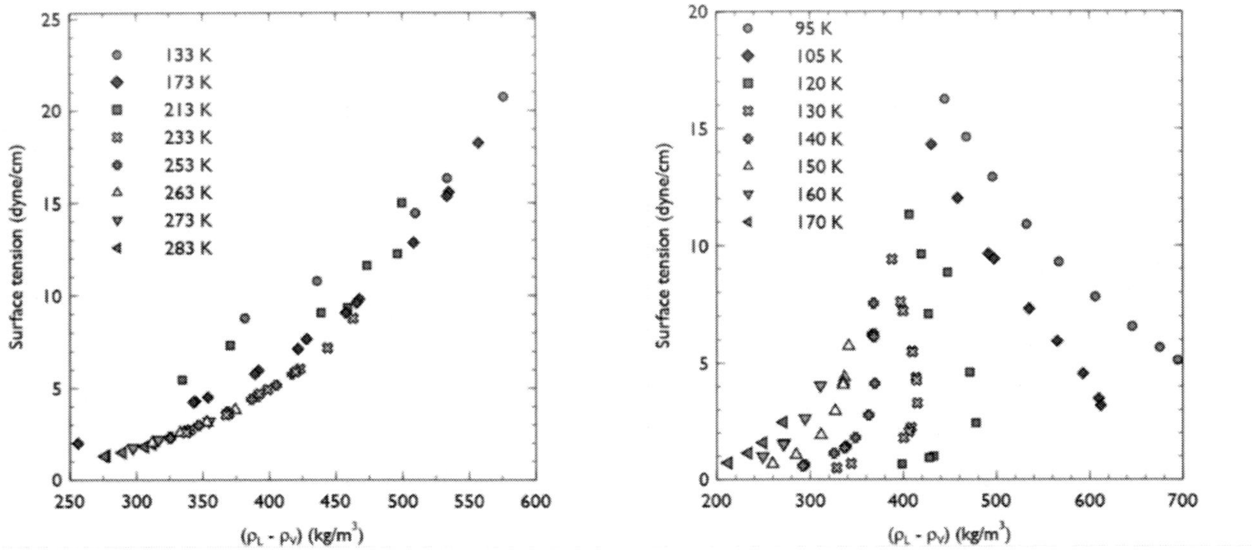

Methane - Ethane   Nitrogen – Methane

Figure 17: Surface tension as a function of density difference for methane – ethane (left) and nitrogen – methane (right). Data from Baidakov et al. (2013, 2012).

We might conclude that one's selected thermodynamic model must accurately represent the densities of both the liquid and vapor phases in order for one to expect an accurate surface tension prediction. The simulations carried out for this work make use of the Peng – Robinson (1976) equation of state (PR EOS). This model is known to perform well in representing vapor-liquid equilibrium (VLE) for many species at low and high pressures, but among its shortcomings is an inaccuracy in the representation of the density of saturated liquid phases as was demonstrated in Fig. 9. This means that, though the PR EOS may be good enough for VLE, it may not be good enough for surface tension.

"New" models for σ

We come quite naturally to a new question:

6. Do we have to accept poor estimates of surface tension?

Or, put another way,

6. Is there anything we can do to improve estimates of surface tension?

What is needed is:

- A model that can be used for all mixtures all of the time

- Failing that, a model that has superior accuracy (compared to most current mixture models) and...
- Can be used when some components are above their own critical points and...
- Is easy to program and rapid in execution (in a computer program)

In the remainder of this paper we consider three other models that, to the best of our knowledge, do not yet feature in any process simulation tool.

Density gradient theory (DGT)

Density gradient theory can hardly be considered new; it dates back in some form to the work of Van der Waals (1894). The approach has the following strengths and weaknesses:

- Is applicable up to true mixture critical point
- Works for any* mixture of chemical species at any conditions
- Permits the use of any EOS/thermodynamic model
- Can be applied to liquid-liquid interfaces as well
- Computationally intensive; hundreds of EOS evaluations per σ calculation
- Requires σ data for all species or accurate c_i model
- Requires binary σ data for cases when $\beta_{ij} = 0$ does not suffice
- Many flavors of DGT algorithms; not all work all of the time
- Requires accurate bulk fluid density estimates
- Nearly always uses a non-cubic equation of state (e.g. a version of SAFT, and there are too many versions of SAFT)
- **Challenging to implement correctly.**

An extensive bibliography of papers pertaining to DGT is provided at the end of this paper. Readers are recommended to refer to this body of literature for the details of the models and methods of calculation and for many examples showing very promising results. Figure 18 shows some results we have obtained with the DGT formulation of Liang et al. (2016, 2017).

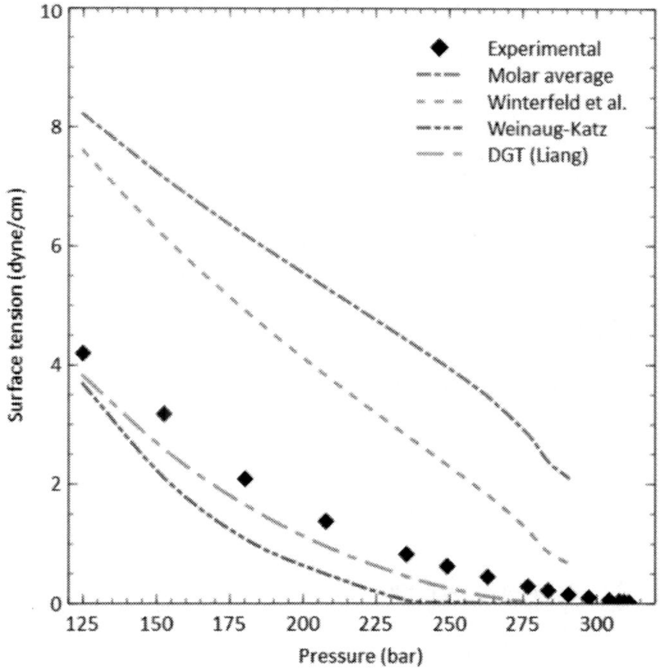

Figure 18: Surface tension as a function of pressure. Data for Fluid A from Danesh et al. (1991).

Cubic EOS (Mulero et al., 2022)

Mulero et al. (2022) have recently developed a cubic equation of state for surface tension:

$$P_c - P = \frac{R^* T\sigma}{M_m - b_m\sigma} - \frac{a_m\sigma^2}{M_m^2 + M_m\sigma(b_m - c_m) - c_m b_m \sigma^2} \quad \textbf{(23)}$$

P_c is the critical pressure and should be estimated from a simple molar average of the pure component values (this detail is omitted from the paper of Mulero et al. which contains several typographical errors.) (Refer to the original paper for an explanation of the other terms in this equation.) The model is computationally trivial but, to date, has been developed only for alkanes. Figure 19 shows some results for the *n*-hexane – *n*-decane mixture. Unfortunately, our results for lower molecular weight alkanes have been much less satisfactory (so far).

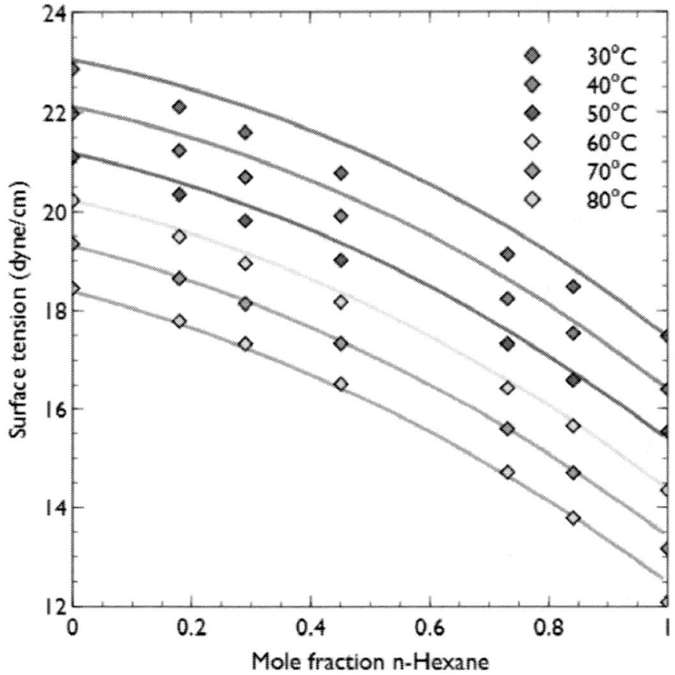

Figure 19: Mulero et al. model for the surface tension of *n*-Hexane – *n*-Decane mixture.

Shardt – Elliott model (2017, 2021)

The Shardt – Elliott model (Shardt & Elliott, 2017; Shardt et al., 2021; Khare et al., 2022; Wang et al, 2022) for surface tension in a binary mixture is:

$$\sigma = \sigma_2 - \left(1 + \frac{bx_2^r}{1 - ax_2^r}\right) x_1^r(\sigma_2 - \sigma_1) \quad \textbf{(24)}$$

Where the reduced mole fractions are defined by:

$$x_1^r = \frac{x_1}{x_1^c} \qquad x_2^r = 1 - x_1^r \quad \textbf{(25)}$$

x_1^c is the mole fraction of species 1 at the critical point (of the mixture), a and b are adjustable parameters.

Figure 20 shows the surface tension for methane – nitrogen; we set $a = 0$ and b tuned only to the top line of data points in order to create this figure.

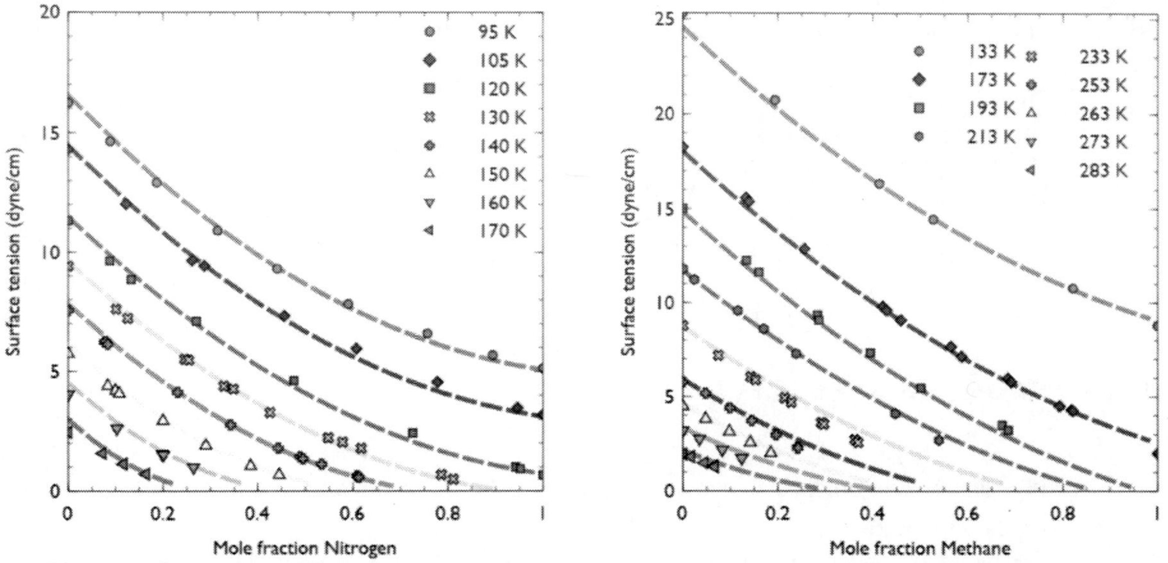

Figure 20: Shardt – Elliott model for the surface tension of a methane – nitrogen mixture (left) and methane – ethane (right).

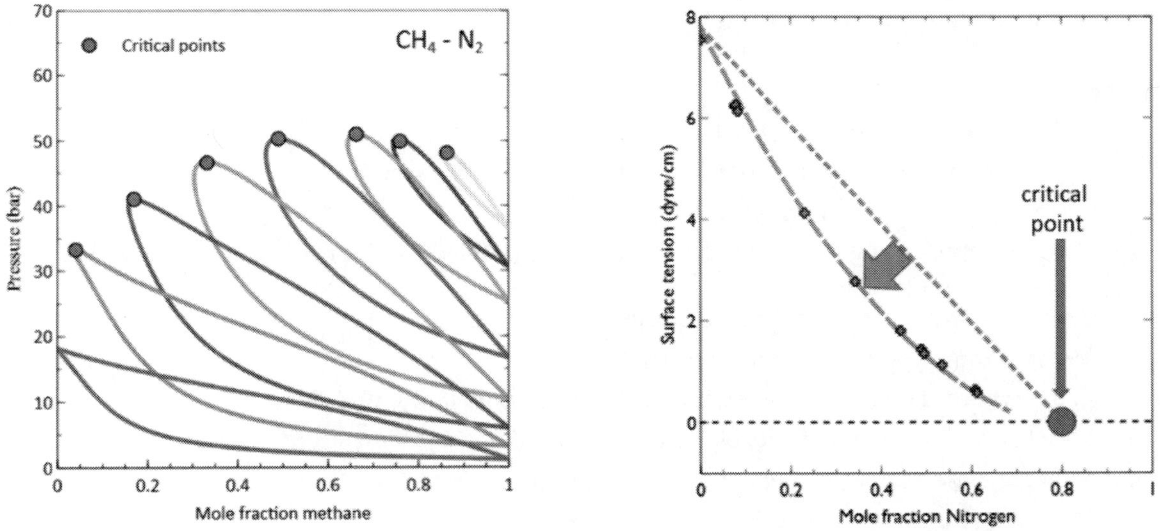

Figure 21: Explanation of how the Shardt-Elliott model works

Figure 21 shows how the model works.

1. Find critical point of mixture at specified temperature. This is a phase equilibrium calculation.
2. At the critical point, $\sigma \rightarrow 0$.
3. Draw straight line from pure component point on left side axis to critical point.
4. Adjust BIPs to move from straight-line approximation to curve that better represents the data.
5. BIPs can be fitted to data at a single temperature. In our examples we kept $a = 0$.

The model for multicomponent systems is rather more complicated:

$$\sigma(x_{i,r},T) = \sum_{i=1}^{n}\left[x_{i,r}(T)\sigma_i(T)\right] + \sum_{i=1}^{n}\left\{\left(\left[\frac{x_{i,r}(T)}{\sum_{j=1}^{n}\left(\frac{x_{j,r}(T)}{1-a_{ji}}\right)}\right]\sum_{j=1}^{n}\left[\frac{b_{ji}}{a_{ji}}\left(\frac{1}{1-a_{ji}}\right)x_{j,r}(T)\left(\sigma_i(T)-\sigma_j(T)\right)\right]\right)\right\} \quad (27)$$

$$x_{i>1,r} = \frac{x_i}{\sum_{j=2}^{n} x_j}\left(1 - x_{1,r}\right) \quad (28)$$

$$x_{1,r}(T) = \frac{x_1}{x_{1,cr}(T)} \quad (29)$$

Conclusion: "Where do we go from here?"

We think that:

1. Currently, there is no surface tension model that is always useful and usable
2. Density gradient theory is
   - Likely to prove the best and most versatile approach
   - May be too time consuming or too difficult to use in routine column simulations
3. The Shardt – Elliott model
   - Appears very promising
   - Is much simpler than DGT
   - Needs BIPs for best results

We plan to...

1. Release a new version of ChemSep that contains
   - The Shardt – Elliott model for multicomponent systems
   - Density gradient theory
2. Pursue some ideas we have for predicting missing Shardt – Elliott BIPs

You, our readers, should ask some more questions

1. What is σ in systems of interest to me?
2. From where did I get those estimates? (Bonus points if measured)
3. Should I believe them? (Probably not)
4. What are the consequences *for me* of incorrect estimates of σ?
5. Pressure your experimentalist friends to measure surface tension in
   - iC_4 – nC_4 mixtures
   - Mixtures of industrial significance
   - Mixtures with more than two components
   - (and to publish the results)
   - So we can develop better models

References

Baidakov, V.G., Kaverin, A.M., Khotienkova, M.N., **2013**. Surface tension of ethane–methane solutions: 1. Experiment and thermodynamic analysis of the results. Fluid Phase Equilibria 356, 90–95. https://doi.org/10.1016/j.fluid.2013.07.008

Baidakov, V.G., Kaverin, A.M., Khotienkova, M.N., Andbaeva, V.N., **2012**. Surface tension of an ethane–nitrogen solution. 1: Experiment and thermodynamic analysis of the results. Fluid Phase Equilibria 328, 13–20. https://doi.org/10.1016/j.fluid.2012.05.005

Billet, R., Schultes, M., **1993**. Predicting mass transfer in packed columns. *Chem. Eng. Technol.* 16, 1–9. https://doi.org/10.1002/ceat.270160102

Billet, R., Schultes, M., **1999**. Prediction of Mass Transfer Columns with Dumped and Arranged Packings: Updated Summary of the Calculation Method of Billet and Schultes. *Transactions of The Institute of Chemical Engineers*. 77, 498–504. https://doi.org/10.1205/026387699526520

Bravo, J.L., **1981**. A Generalized Correlation for Mass Transfer in Packed Distillation Columns (Master of Science in Engineering). University of Texas at Austin, Austin, Texas.

Bravo, J.L., Fair, J.R., **1982**. Generalized correlation for mass transfer in packed distillation columns. *Ind. Eng. Chem. Proc. Des. Dev.* 21, 162–170. https://doi.org/10.1021/i200016a028

Brock, J.R., Bird, R.B., **1955**. Surface tension and the principle of corresponding states. *AIChE J.* 1(2), 174–177. https://doi.org/10.1002/aic.690010208

Boudh-Hir, M.E., Mansoori, G.A., **1990**. Statistical mechanics basis of Macleod's formula. *J. Phys. Chem.* 94(21), 8362–8364. https://doi.org/10.1021/j100384a068

Danesh, A. S., Dandekar, A. Y, Todd, A. C., Sarkar, R. A Modified Scaling Law and Parachor Method Approach for Improved Prediction of Interfacial Tension of Gas-Condensate Systems. Presented at the *SPE Annual Technical Conference and Exhibition*, Dallas, Texas, October 1991. Paper SPE-22710-MS.

Gupta, S., Elliott, J.R., Anderko, A., Crosthwaite, J., Chapman, W.G., Lira, C.T., **2023**. Current Practices and Continuing Needs in Thermophysical Properties for the Chemical Industry. *Ind. Eng. Chem. Res.* 62, 3394–3427. https://doi.org/10.1021/acs.iecr.2c03153

Horstmann, S., Grybat, A., Ihmels, C., Fischer, K., **2010**. High Pressure Demethanizer Physical Properties (Research Report No. RR-203). Gas Processors Association.

Khare, A., Shardt, N., Elliott, J.A.W., **2022**. Accuracy of the Shardt–Elliott–Connors–Wright equation for predicting the surface tension of metal alloys. Journal of Alloys and Compounds 922, 166117. https://doi.org/10.1016/j.jallcom.2022.166117

Lee, S.T., Chien, M.C.H., **1984**. A New Multicomponent Surface Tension Correlation Based on Scaling Theory, in: SPE Enhanced Oil Recovery Symposium. Presented at the SPE Enhanced Oil Recovery Symposium, Society of Petroleum Engineers, Tulsa, Oklahoma. https://doi.org/10.2118/12643-MS

McCarley, K., December **2021**. Finding the Capacity of a Distillation Column. *CEP*. 23–28.

Macleod, D.B., **1923**. On a relation between surface tension and density. *Trans. Faraday Soc.* 19, 38. https://doi.org/10.1039/TF9231900038

Mulero, A., Cachadiña, I., Cardona, L.F., Valderrama, J.O., **2022**. Pressure–Surface Tension–Temperature Equation of State for n-Alkanes. Ind. Eng. Chem. Res. 61, 3457–3473. https://doi.org/10.1021/acs.iecr.1c04979

Shardt, N., Elliott, J.A.W., **2017**. Model for the Surface Tension of Dilute and Concentrated Binary Aqueous Mixtures as a Function of Composition and Temperature. Langmuir 33, 11077–11085. https://doi.org/10.1021/acs.langmuir.7b02587

Shardt, N., Wang, Y., Jin, Z., Elliott, J.A.W., **2021**. Surface tension as a function of temperature and composition for a broad range of mixtures. Chemical Engineering Science 230, 116095. https://doi.org/10.1016/j.ces.2020.116095

Stupin, W.J., Kister, H.Z., **2003**. System Limit: The Ultimate Capacity of Fractionators. *Chem. Eng. Res. Des.* 81, 136–146. https://doi.org/10.1205/026387603321158294

Summers, D.R., **2023**. Practical Industrial Distillation, Paper presented at the AIChE Spring Meeting, March 13, 2023, Houston, Texas

Sugden, S., **1924**. CXLII.—A relation between surface tension, density, and chemical composition. *J. Chem. Soc., Trans.* 125, 1177–1189. https://doi.org/10.1039/TF9231900038

Tyn, M.T., Calus, W.F., **1975**. Diffusion coefficients in dilute binary liquid mixtures. Journal of Chemical and Engineering Data 20, 106–109. https://doi.org/10.1021/je60064a006

Wang, Y., Shardt, N., Elliott, J.A.W., Jin, Z., **2022**. Highly Efficient and Accurate Gas-Alkane Binary Mixture Interfacial Tension Equations for a Broad Range of Temperatures, Pressures, and Compositions. SPE Journal 27, 895–913. https://doi.org/10.2118/208572-PA

Weinaug, C.F., Katz, D.L., **1943**. Surface Tensions of Methane-Propane Mixtures. Ind. Eng. Chem. 35, 239–246. https://doi.org/10.1021/ie50398a028

Winterfeld, P.H., Scriven, L.E., Davis, H.T., **1978**. An approximate theory of interfacial tensions of multicomponent systems: Applications to binary liquid-vapor tensions. *AIChE J.* 24(6), 1010–1014. https://doi.org/10.1002/aic.690240610

References pertaining to Density Gradient Theory

There is a very significant body of work on DGT. Here is a list (not exhaustive) of papers on the topic.

Breure, B., Peters, C.J., 2012. Modeling of the surface tension of pure components and mixtures using the density gradient theory combined with a theoretically derived influence parameter correlation. Fluid Phase Equilibria 334, 189–196. https://doi.org/10.1016/j.fluid.2012.08.010

Cachadiña, I., Hernández, A., Mulero, Á., 2022. Surface tension of esters. Temperature dependence of the influence parameter in density gradient theory with Peng-Robinson equation of state. Case Studies in Thermal Engineering 36, 102193. https://doi.org/10.1016/j.csite.2022.102193

Cea-Klapp, E., Gajardo-Parra, N., Aravena, P., Quinteros-Lama, H., Held, C., Canales, R.I., Garrido, J.M., 2022. Interfacial Properties of Deep Eutectic Solvents by Density Gradient Theory. Ind. Eng. Chem. Res. 61, 2580–2591. https://doi.org/10.1021/acs.iecr.1c03817

Enders, S., Quitzsch, K., 1998. Calculation of Interfacial Properties of Demixed Fluids Using Density Gradient Theory. Langmuir 14, 4606–4614. https://doi.org/10.1021/la9712707

Haarmann, N., Reinhardt, A., Danzer, A., Sadowski, G., Enders, S., 2020. Modeling of Interfacial Tensions of Long-Chain Molecules and Related Mixtures Using Perturbed Chain-Statistical Associating Fluid Theory and the Density Gradient Theory. J. Chem. Eng. Data 65, 1005–1018. https://doi.org/10.1021/acs.jced.9b00339

Hosseini, S.M., Sadeghi, M., Zarei, T., 2022. A combined density gradient theory with equation of state model for the study of surface tension of refrigerant fluids. Journal of Molecular Liquids 352, 118629. https://doi.org/10.1016/j.molliq.2022.118629

Kahl, H., Enders, S., 2000. Calculation of surface properties of pure fluids using density gradient theory and SAFT-EOS. Fluid Phase Equilibria 172, 27–42. https://doi.org/10.1016/S0378-3812(00)00361-7

Kahl, H., Enders, S., 2002. Interfacial properties of binary mixtures. Physical Chemistry Chemical Physics 4, 931–936. https://doi.org/10.1039/b108535m

Larsen, P.M., Maribo-Mogensen, B., Kontogeorgis, G.M., 2016. A collocation method for surface tension calculations with the density gradient theory. Fluid Phase Equilibria 408, 170–179. https://doi.org/10.1016/j.fluid.2015.08.024

Liang, X., Michelsen, M.L., 2017. General approach for solving the density gradient theory in the interfacial tension calculations. Fluid Phase Equilibria 451, 79–90. https://doi.org/10.1016/j.fluid.2017.07.021

Liang, X., Michelsen, M.L., Kontogeorgis, G.M., 2016a. A density gradient theory based method for surface tension calculations. Fluid Phase Equilibria 428, 153–163. https://doi.org/10.1016/j.fluid.2016.06.017

Liang, X., Michelsen, M.L., Kontogeorgis, G.M., 2016b. Pitfalls of using the geometric-mean combining rule in the density gradient theory. Fluid Phase Equilibria 415, 75–83. https://doi.org/10.1016/j.fluid.2016.01.047

Mairhofer, J., Gross, J., 2018. Modeling properties of the one-dimensional vapor-liquid interface: Application of classical density functional and density gradient theory. Fluid Phase Equilibria 458, 243–252. https://doi.org/10.1016/j.fluid.2017.11.032

Mairhofer, J., Gross, J., 2017. Modeling of interfacial properties of multicomponent systems using density gradient theory and PCP-SAFT. Fluid Phase Equilibria 439, 31–42. https://doi.org/10.1016/j.fluid.2017.02.009

Moslehi, H., Hosseini, S.M., Alavianmehr, M.M., 2023. Density gradient theory study of surface tension of pure and mixture of refrigerant fluids with the help of perturbed-hard-chain equation of state. Fluid Phase Equilibria 568, 113751. https://doi.org/10.1016/j.fluid.2023.113751

Mousazadeh, M.H., Faramarzi, E., 2011. Calculation of surface tension of metals using density gradient theory and PC-SAFT equation of state. Fluid Phase Equilibria 301, 13–17. https://doi.org/10.1016/j.fluid.2010.11.006

Mu, X., Frank, F., Alpak, F.O., Chapman, W.G., 2017. Stabilized density gradient theory algorithm for modeling interfacial properties of pure and mixed systems. Fluid Phase Equilibria 435, 118–130. https://doi.org/10.1016/j.fluid.2016.11.024

Mu, X., Frank, F., Riviere, B., Alpak, F.O., Chapman, W.G., 2018a. Mass-Conserved Density Gradient Theory Model for Nucleation Process. Ind. Eng. Chem. Res. 57, 16476–16485. https://doi.org/10.1021/acs.iecr.8b03389

Mu, X., Xi, S., Alpak, F.O., Chapman, W.G., 2018b. Modified Density Gradient Theory for Surfactant Molecules Applied to Oil/Water Interfaces. Ind. Eng. Chem. Res. 57, 7643–7654. https://doi.org/10.1021/acs.iecr.8b00164

Niño Amézquita, O.G., Enders, S., Jaeger, P.T., Eggers, R., 2010. Interfacial properties of mixtures containing supercritical gases. The Journal of Supercritical Fluids 55, 724–734. https://doi.org/10.1016/j.supflu.2010.09.040

Rehner, P., Gross, J., 2018. Predictive density gradient theory based on nonlocal density functional theory. Phys. Rev. E 98, 063312. https://doi.org/10.1103/PhysRevE.98.063312

Schäfer, E., Sadowski, G., Enders, S., 2014. Interfacial tension of binary mixtures exhibiting azeotropic behavior: Measurement and modeling with PCP-SAFT combined with Density Gradient Theory. Fluid Phase Equilibria 362, 151–162. https://doi.org/10.1016/j.fluid.2013.09.042

Shen, G., Held, C., Lu, X., Ji, X., 2016. Modelling interfacial properties of ionic liquids with ePC-SAFT combined with density gradient theory. Molecular Physics 114, 2492–2499. https://doi.org/10.1080/00268976.2016.1171408

Stephan, S., Becker, S., Langenbach, K., Hasse, H., 2020. Vapor-liquid interfacial properties of the system cyclohexane + CO2: Experiments, molecular simulation and density gradient theory. Fluid Phase Equilibria 518, 112583. https://doi.org/10.1016/j.fluid.2020.112583

Stephan, S., Langenbach, K., Bart, H.-J., 2018a. Enrichment of components at vapour - liquid interfaces: a study by molecular simulation and density gradient theory. Chemical Engineering Transactions 69, 295–300. https://doi.org/10.3303/CET1869050

Stephan, S., Langenbach, K., Hasse, H., 2019. Interfacial properties of binary Lennard-Jones mixtures by molecular simulation and density gradient theory. J. Chem. Phys. 150, 174704. https://doi.org/10.1063/1.5093603

Stephan, S., Liu, J., Langenbach, K., Chapman, W.G., Hasse, H., 2018b. Vapor−Liquid Interface of the Lennard-Jones Truncated and Shifted Fluid: Comparison of Molecular Simulation, Density Gradient Theory, and Density Functional Theory. J. Phys. Chem. C 122, 24705–24715. https://doi.org/10.1021/acs.jpcc.8b06332

Sun, Y., Zuo, Z., Shen, G., Held, C., Lu, X., Ji, X., 2021. Modeling interfacial properties of ionic liquids with ePC-SAFT combined with density gradient theory. Fluid Phase Equilibria 536, 112984. https://doi.org/10.1016/j.fluid.2021.112984

Van der Waals, J.D., **1894**. Thermodynamische theorie der kapillarität unter voraussetzung stetiger dichteänderung. *Z. Phys. Chem.* 13, 657−725. https://doi.org/10.1515/zpch-1894-1338 (English translation: Rowlinson, J.S., **1979**. Translation of J.D. van der Waals' "The thermodynamic theory of capillarity under the hypothesis of a continuous variation of density". *J. Statist. Phys.* 20, 197−244. https://doi.org/10.1007/BF01011513).

Vinš, V., Aminian, A., Celný, D., Součková, M., Klomfar, J., Čenský, M., Prokopová, O., 2021. Surface tension and density of dielectric heat transfer fluids of HFE type-experimental data at 0.1 MPa and modeling with PC-SAFT equation of state and density gradient theory. International Journal of Refrigeration 131, 956–969. https://doi.org/10.1016/j.ijrefrig.2021.06.029

Vinš, V., Planková, B., Hrubý, J., 2013. Surface Tension of Binary Mixtures Including Polar Components Modeled by the Density Gradient Theory Combined with the PC-SAFT Equation of State. Int J Thermophys 34, 792–812. https://doi.org/10.1007/s10765-012-1207-z

Wendt, R.P., 1962. The Density Gradient and Gravitational Stability During Free Diffusion in Three-Component Systems. J. Phys. Chem. 66, 1740–1742. https://doi.org/10.1021/j100815a505

125e Experience with Using Group Contribution Methods for Phase Equilibria in Industrial Distillation Column Design

Umesh Mathur

It is common to find mixtures with many (10-20) compounds in many, if not most complex chemical separation processes. Activity coefficient models are imperative when modeling the phase equilibria for such systems (vapor-liquid, vapor-liquid-liquid, or liquid-liquid). This effort requires providing binary interaction parameters for each constituent binary in the mixture. For example, with 15 components, one would need to characterize the behavior of 105 binaries.

Ideally, one should find experimentally measured phase equilibrium data for each such binary and regress the activity coefficient parameters for the chosen activity coefficient model. In attempting to do so, there are, unfortunately, several immediate issues to be confronted: (a) such data are scattered throughout the world's literature (often in foreign language journals), (b) it is generally impractical to attempt to find them in a reasonable amount of time, (c) there is no guarantee that published data are in fact at all reliable – in fact, there are many published datasets that are "manufactured' or completely fake, and, finally, (d) a random or careless search could well fail to find data that do exist somewhere.

Alternatively, one could access computerized databanks, developed over the years, for storing such data that have been screened by domain experts and made available in an organized way; however, these databanks are not inexpensive. Besides, it may well happen that no such data exist for several binaries in question. In such situations, group contribution methods (GCMs) would be used to estimate phase equilibria for such binaries, provided they are not 'key components' in the separation system. If the binaries in question do, in fact, represent such key components, there would be no alternative but to insist on commissioning experimental phase equilibria measurements from a qualified laboratory.

These GCM methods originated many decades ago and have been updated over time. Regrettably, the group interaction parameters for the most reliable GCMs are not in the public domain and require licensing arrangements to access the latest developments.

The major process simulators used in the chemical industry do allow users to provide their own GCM interaction parameters; however, they are not allowed to provide them by default. In addition, while they provide built-in activity coefficient parameters for selected binary mixtures, it is often the case that these are not properly documented as to their sources:

- Were these parameters regressed using measured phase equilibria?
- Were all available data for each binary mixture used, or was some subset used naively without regard to limitations on extrapolation way beyond the range of the data (e.g., for pressure or temperature)?
- Were these parameters estimated using some GCM based on outdated or inaccurate GCM interaction parameters?

This paper will present examples of a few successful industrial projects where the issues outlined above were addressed reliably. The discussion will show how practitioners can avoid serious (but subtle) flaws in the thermodynamic models that lie behind every industrial process simulation. Such errors can lead to disastrous designs that fail to meet throughput or product quality guarantees. There are no easy shortcuts, and it behooves all practitioners to be mindful of, and plan for, the non-trivial effort required to ensure that their designs are successful.

125f Impact of Surface Tension in Distillation. Part II: Impact of Surface Tension on High Pressure Column Design

Ruan de Villiers, Attilio J. Praderio, Hendrik A. Kooijman, Ross Taylor

Surface tension is critical in determining bubble and droplet dynamics in trayed and packed distillation columns. Fair's well known correlation for flooding in tray columns employs a relative surface tension raised to the power of 0.2 with the reference value of the surface tension equal to 20 dynes/cm. The Billet and Schultes model for packed column performance sets a lower limit of 30 dynes/cm. The low (or non-existent) power dependency meant that engineers did not worry much about errors of a few dyne/cm.

In some high-pressure columns (demethanizers and nitrogen rejection units come to mind) the surface tension may become quite low, close to or even lower than 1 dyne/cm. When the surface tension is that low, errors of a few dynes/cm can easily alter the column diameter (or height) by a significant amount. Is the actual surface tension really that low? The answer could well be yes; we do know that the surface tension reaches zero at the critical point.

In simulation and design of these types of columns engineers almost inevitably resort to estimating the surface tension using a model that is available in the process simulation tool in use. Since surface tension is, arguably, the least well predicted physical property, often it is estimated by means of the simplest mixture methods available. In many cases the only choice of estimation model is a mole fraction weighted average of the pure component surface tensions. Sometimes, this is the worst possible model, but many of the alternatives are not much better (and can be worse in some circumstances). The mole fraction average is particularly problematic when one or more of the components is present at conditions above their own individual critical points (but the mixture as a whole is well below its own).

What is needed is a high fidelity model for mixture surface tensions that:

1. Retains accuracy at low surface tensions (close to mixture critical points),
2. Can handle mixtures in which some components are above their own critical points,
3. Does not cause column performance models (mass-transfer coefficients, interfacial areas, and pressure drop) to behave inappropriately when the surface tension is low (while not employing arbitrary cut-off values of the surface tension), and
4. Behaves reliably *and* efficiently when used in a column design simulation.

Models that claim to address at least some of these issues now exist; they come in three classes:

1. Density Gradient Theory (DGT)
2. Cubic models (analogous to cubic equations of state)
3. The Shardt-Elliot type of model

As far as we know, none of these models has yet been implemented in process design software.

The objective of Part II is to use these models in column simulations (we use the ChemSep package) in order to demonstrate the extent to which an inaccurate surface tension can impact column simulation and design.

125g Quantifying the Effect of Stripping Factor on Distillation Efficiency Using Cyclohexane/n- Heptane Test System

Rukhsar Ahmed, Tony Cai, Izak Nieuwoudt, Sayeed Mohammad, Clint Aichele

Distillation is one of the oldest separation processes and the most widely used unit operation in the chemical process industries. About 95% of all separations are accomplished with distillation. Since distillation is also an energy-intensive process, even small improvements in distillation efficiency would have a significant impact on the design and operation of distillation columns.

Although several experimental variables that affect the distillation tray efficiency have been explored in the past, nearly all of the previous studies were conducted under total reflux conditions. However, a non-total reflux condition more closely represents the actual operation of a commercial distillation column. Therefore, in this study, we present recent results obtained from an Oldershaw column under non-total reflux conditions. In particular, an Oldershaw column was designed, installed and commissioned that is capable of operating under total as well as non-total reflux conditions. The validation was performed using data for cyclohexane/n-hexane test system from earlier work as well as FRI data on sieve trays.

Since the stripping factor is an important variable for investigating the distillation efficiency under non-total reflux conditions, a systematic study was conducted using the cyclohexane/n-hexane test system over a range of stripping factor values and column pressures from vacuum to atmospheric conditions. Results to date indicate that the point efficiency of the rectifying section was greater than that of the stripping section with the measured point efficiencies falling in the range of 40%-60%. The experimental results were also compared with the L/V ratios and the product concentrations obtained from column simulation in Aspen Plus®, both of which showed close agreement. Important findings from this ongoing study will be discussed.

125h Converging Distillation Tower, Design, and Optimization

Avinashkumar Karre, Abhishek Jayswal

The distillation columns are used globally in the industries. Some columns are simple with two compounds and several columns are complex in nature processing multicomponent system. The presentation will explain theory behind the design of the distillation tower using K value and relative volatility. Effect and design considerations of several process parameters such as pressure, reflux ratio, pump around requirement etc. on the design of distillation column is explained. Stepwise tower optimization techniques are illustrated. A procedure to simulation a tower using a process simulator is discussed. Setting up a complex column using a process simulation and steps to converge the simulation is discussed.

147a Common Mistakes in Distillation Control and Their Remedies

Michael Taube, Keith Lehman, Isuru A. Udugama

Distillation is involved in every one of the factions of the Process Industries (PI), from Oil & Gas, through Refining, Petrochemicals and Specialty Chemicals, as well as the bio-process industry. In spite of the very long history of design and operations of these unit operations, there is a recurring theme of misapplied process & mechanical designs and sub-optimal control structures which result in poor and/or under-performing process units.

This paper reviews many of the common mistakes encountered by the authors and the remedies which address them.

147c Best Practices in Column Control Distilled

Mark Darby, Doug Nicholson, Greg McMillan

The authors recently combined efforts to cover distillation column control in a series of articles*. This presentation summarizes our views on what we believe represent best practices for distillation column control. Topics covered include selecting the regulatory control structure based on the given objectives and typical disturbances, selecting the best tray for temperature or soft sensor control, and when dual-ended composition control is warranted.

Also discussed are algorithm selection, PID tuning considerations, managing constraints, and when to consider Model Predictive Ccontrol (MPC). Guidance is also provided on the role of simulation and when to consider optimization.

Examples drawn from experience are provided to reinforce the recommendations.

*'Better Column Control Distilled – Part 1(-5)" in Control Talk, McMillan, G., *Control Global*, July (-December) 2022.

Develop Distillation Controls during Process Design Phase

Charles D. Herzog

Retired Chemical Engineer

Address basic process control issues during engineering design phase to increase system capacity and improve on-spec performance.

View process control as a life-cycle discipline, rather than an activity that begins after startup.

Summary:

This paper is a follow-up to a February 2021 CEP paper [1] discussing process design improvements that benefit distillation process control. This paper summarizes the 2021 publication and discusses additional process design upgrades.

The 2021 paper recommends these improvements to reduce process variation, increase on-spec capacity, and reduce energy consumption:

- Insulating overhead and feed systems
- Increasing residence time for products and feed drum
- Adding redundant analyzer for primary product
- Designing process control strategy to prioritize primary product

By reducing variation, the frequency and magnitude of off-spec incidents are decreased.

Some of the focus areas for this paper are:

- Calculating equipment loads to achieve target purity rather than product spec
- Using a specific process control strategy to justify instruments and analyzers
- Designing analyzer sample systems to preserve sample integrity
- Designing light ends vent systems to minimize pressure variation and product loss

Introduction:

Process control engineering on new distillation systems is usually deferred until after startup. P&ID's are populated with the 'standard' process control strategy (Fig 3), in order to complete the specification and procurement of instruments. Minimizing capital cost and staying on schedule are of primary importance to design teams.

Basic regulatory control strategies are rarely changed after startup, or even after HAZOP. Operator training materials would have to be revised and operators re-trained. HAZOP would have to be repeated. Thus the standard process control strategy usually becomes permanent.

Process control engineering activities during design phase fall into three main subgroups:

- Equipment and piping design
- Process control strategies
- Process design of analyzer applications

Each of these subgroups is discussed in detail below.

Discussion:

Design for Changing Conditions and Constrained Equipment

Distillation systems are usually designed for steady-state conditions. Basis of design documents refer to a maximum and minimum feed rate, and a range of feed compositions and ambient temperature conditions. Separate cases representing the upper and lower limits are simulated, and the controlling load case is selected for equipment design.

Rarely, if ever, is the unit designed for changing conditions, despite the fact that changes in feed rate, composition, and ambient conditions can all cause off-spec production.

Additionally, the regulatory control system must be designed to keep the primary product on-spec when equipment constraints are active, and to quickly restore heat balance following a disturbance.

Select an Operating Target for Design Simulation

The design process begins by selecting a simulation operating target with higher purity than the product spec. Equipment loads should be calculated so that constraints do not cause unsteady operation at design rates.

The target composition should be determined based on the predicted composition variation of the product (Fig 2a, 2b). One way to predict variation is to study operating data from similar columns. Variation can be reduced by implementing design improvements listed below.

Equipment and Piping Design:

Fig 1 illustrates a typical distillation column. As discussed in the 2021 paper, equipment and piping upgrades to reduce variation include:

- Insulate feed and overhead systems to reduce pressure loss caused by rainstorms
- Increase residence time for each product to reduce variation of composition
- Increase residence time for feed drum to reduce variations in feed flow and composition

Of great importance is that increasing the reflux drum residence time not only reduces variation of the distillate composition, but also reduces variation of reflux composition. This reduces variation of liquid composition on every tray, benefiting all objectives of distillation.

An additional benefit of increasing product holdup is to give the operator more time to respond to feed composition changes and maintain on-spec production. An example calculation is given in the 2021 paper.

Light Ends Vent System

An underrated subject for process design teams is the vent system. All systems with a total liquid distillate product require a vent system for removal of light ends and nitrogen. Without a vent system the light ends accumulate, decreasing the partial pressure of distillate components, reducing condenser LMTD and capacity.

An undesirable consequence of venting is loss (or recycle) of distillate product in the vent stream. The design team must address the following:

- Optimum destination for vent (feed drum is the default)
- Venting intermittently vs continuously
- Vent line size and controls
- Use of a vent condenser to reduce product loss

The vent destination is site specific. If there is no continuous source of light ends entering the system, then the project can choose either intermittent or continuous venting. Intermittent venting reduces product loss, but causes a pressure disturbance on the column every time the vent is opened or closed. At high rates, venting may be required more frequently to relieve constraints caused by buildup of light ends.

Continuous venting reduces pressure variation and prevents buildup of light ends, but reduces net capacity of the entire distillation system. Continuous venting options should be simulated and fully designed.

A simple vent condenser can be a jacketed section of vertical pipe above the reflux drum. Otherwise a small vertical heat exchanger can be used. Any condensed material falls back to the reflux drum by gravity. A restriction orifice can be used to limit the flow. Vent streams from most light hydrocarbon towers are clean and unlikely to cause plugging.

Process Control Strategies:

The standard process control strategy (Fig 3) is automatically chosen on most distillation projects. With 'fast-track' projects, any mention of a different process control strategy is quickly silenced. Most process design engineers lack experience with process control and show the standard strategy on the P&ID without a second thought.

It is up to senior management at design firms to recognize process control as a required skill area, and to provide basic process control training. Trying to design a distillation system without process control experience is like trying to design an automobile without having ever driven one.

Selection of Distillation Process Control Strategy

The strategy must be evaluated according to which product, either distillate or bottoms, is primary. An example of a tower with a primary distillate product is a C3 splitter. A tower with a primary bottoms product is a debutanizer in a fluid catalytic cracking (FCC) unit. Both of these applications are discussed.

The place to begin distillation control engineering is to list the independent and dependent variables. Table 1 lists the pairing of variables in the standard process control strategy.

Independent Variables	Dependent Variables
Condenser Area or Fans	Overhead Pressure
Distillate Product Flow Rate	Accumulator Level
Bottoms Product Flow Rate	Bottoms Level
Reflux Flow Rate	Distillate Product Composition
Reboiler Heat Duty	Bottoms Product Composition

Table 1: Independent and Dependent Distillation Process Variables for Standard Strategy

When equipment constraints are not active, the system has zero degrees of freedom, and compositions of both products can be controlled. The standard strategy succeeds in this case.

When the first of the independent variables in Table 1 becomes constrained, the control system has negative degrees of freedom, and it is no longer possible to control composition at both ends. The standard control strategy (Fig 3) always prioritizes the composition of the bottoms product via temperature control. This has the net effect of sacrificing distillate purity on all columns when constraints become active, an unacceptable outcome.

Constrained Condenser Causes Off-Spec Operation and Flaring

Fig 4 represents an actual example of a capacity test run that demonstrated the shortcomings of the standard process control strategy.

The distillate is the primary product with a 98% purity spec. Pressure is controlled by active condenser area.

As shown in Fig 4, the standard process control strategy is used. Operator increases purity of the distillate by increasing reflux. The tower is heat balanced by a bottoms temperature controller, and each level is controlled by its own product flow rate.

Operator increases the feed rate to begin the formal capacity test. At first, operator successfully increases reflux to keep distillate on-spec. Bottoms tray temperature is at the normal setting and reasonably steady.

As the day gets warmer, cooling water supply temperature increases and the condenser shows signs of becoming constrained. Operator increases reflux in an attempt to control distillate purity.

The bottoms temperature controller sees the increased reflux and increases reboiler heat duty to keep the temperature at its set point. This increases vapor flow to the constrained condenser, increasing pressure to the point where the vent to flare opens, causing the test to fail (Fig 5).

In fact, the only way for the operator to keep the distillate on-spec in this scenario is to use the bottoms temperature set point to manage the constraint. This is a difficult, if not impossible task, because of dead time associated with temperature control. During the actual test, neither the operator nor support engineers even recognized the active constraint until flaring began.

Is There a Better Strategy?

After the failed performance test, the DCS control strategy was revised as shown in Fig 6. The test was successful with the new strategy.

The revised strategy is designed specifically for the constrained environment with negative degrees of freedom. The bottoms temperature control is discarded and bottoms composition is allowed to float when constraints are active.

Primary distillate product composition is controlled by material balance rather than by reflux. Increasing the bottoms product rate removes heavy ends from the system. The required change in bottoms rate can be directly calculated either by operator or by a simple DCS application.

The system is heat balanced by bottoms level control, which reaches steady state much faster than temperature control. This reduces vapor flow up the column, thereby improving purity of the distillate and reducing condenser load.

The net result is to improve the product purity by taking load off the system. Thus composition of primary product is always controlled, even when a constraint is active.

FCC Debutanizer and the Standard Process Control Strategy

The bottoms gasoline is the primary product in the FCC debutanizer. The standard strategy is used in most, if not all, FCC debutanizers. The bottoms temperature controller seems to be in the correct place for this tower, controlling the C4 concentration (RVP) in the gasoline.

However, the bottoms temperature controller is overburdened. It is responsible both for bottoms composition and for the heat balance. Whenever the temperature is constant, the column is in heat balance, but not necessarily at the target gasoline composition. The temperature set point has to be changed to achieve target composition. Each change in set point takes the column out of heat balance for an hour or more, because of long response time associated with tray temperature control.

Fig 8 shows the preferred debutanizer control strategy as a mirror image of the C3 splitter. The gasoline RVP is controlled by adjusting C4 distillate product flow rate, and the column is heat balanced by reflux drum level control. Heat balance by level control has a much faster response than heat balance by temperature control. Reboiler vapor is adjusted to manage constraints.

Using the Process Control Strategy to Justify Additional Instruments or Analyzers

Fig 6 shows a C3 splitter with the material balance process control strategy. Bottoms product rate is manipulated to control distillate purity.

In this example, if concentration of propane in the feed stream increases, the bottoms flow rate must be increased to keep the distillate on-spec. According to the strategy of Fig 6, the operator would increase bottoms flow when the distillate shows increasing propane concentration. This brings the propylene product back to the target, but composition of the reflux / product material has already changed. Even after the operator intervenes, the distillate purity takes an hour or more to return to target.

However, if a feed analyzer is added (Fig 9), operator can begin to increase bottoms flow much sooner, reducing or eliminating variation of the distillate product. In fact, a propane balance model can be created to directly calculate the required change of bottoms flow rate in response to changes in propane concentration of the feed. A response based on propane balance brings the bottoms flow to the correct place soon enough to keep distillate purity constant.

In this case the propane balance control strategy justifies addition of a feed analyzer to the project. The ability to demonstrate precisely how the proposed new analyzer will be used greatly increases the chance of convincing the project team of its value.

Analyzer Applications and Sample System Design Issues

There are two major improvement areas for process design teams when it comes to analyzer systems:

- Adding redundant analyzers for products
- Designing sample systems to ensure integrity of the sample

The issue of redundant analyzers was discussed in the 2021 paper. Without a redundant analyzer, routine calibrations leave the operator in the dark for the entire calibration period. Operators typically increase reflux to ensure on-spec production when the analyzer goes out of service. This moves the system to a less efficient operation. Moving the system back to the target requires significant operator attention. The process may take 24 hours or more.

Adding a redundant analyzer for the primary product completely eliminates this problem, and merits serious consideration on every project. Note that the redundant analyzer costs far less than the original because it is a duplicate.

Designing Analyzer Sample Systems to Preserve Sample Integrity

Analyzer sample systems are process lines, usually ½" tubes, connecting the process to the analyzer building (Fig 12). Sample lines can be hundreds of feet long. The purpose of the sample system is to deliver an unchanged sample to the analyzer. This usually means maintaining a single phase over the entire length of the sample line. The samples requiring the greatest engineering attention are:

- Saturated vapor samples that are warmer than winter ambient temperature
- Saturated liquid samples that are cooler than summer ambient temperature

Figure 11 shows the case of a critical saturated vapor sample being drawn at a temperature above ambient. Liquids rich in the heavy key component condense and accumulate in low point pockets along the route (Fig 12). The vapor going to the analyzer building has a lower concentration of the heavy key component than the original sample. Operator may not be aware that this is happening. This phenomenon may coincide with the first cold front of the fall.

As condensation continues, a liquid seal forms. This prevents vapor from flowing in the sample line and forces a slug of heavy liquid into the analyzer. The analyzer then shows an unreasonably high concentration of heavy key component, possibly beyond the analyzer's high range value.

At a minimum, sample lines vulnerable to condensation must at least be insulated, or even heated by contact with a warm process line such as cooling water return. No low point pockets should be specified for vulnerable vapor samples, and no high point pockets for vulnerable liquid samples.

An example of a saturated liquid sample below summer ambient temperature is the bottoms stream from an ethylene plant demethanizer (50°F). Without insulation, vaporized material may accumulate in the high points of the sample line during summer. Analysis of this stream has proven so difficult that the methane analysis is moved to deethanizer overhead vapor.

Startup Example:

The condensation phenomenon occurred at a plant startup (Fig 11). Ambient temperature was about 35°F. The saturated vapor distillate (at 100°F) from a light hydrocarbon column feeds directly to an exothermic reactor. The concentration of heavy key component, a di-olefin, is measured by a gas chromatograph (GC) analyzer. Excess quantities of the di-olefin can lead to a runaway exothermic reaction. Operator diverts the reactor feed stream to flare when necessary.

The distillation system had been running for a few hours and seemed to be working correctly. Flow rates and temperature profile matched the data sheet reasonably well. Operators had been diverting distillate vapor to flare, awaiting permission to start the reactor.

However, the distillate analyzer indicated high concentration of the heavy key, beyond the top of scale, and the operations manager was hesitant to start the reactor. Most of the personnel suspected that the analyzer was malfunctioning, but nobody knew why.

The condensation phenomenon can be difficult to identify. Checking the analyzer does not help because the analyzer works fine when being tested on a calibration standard. The instrument engineer proved the condensation mechanism by disconnecting the sample tubing and purging liquid out with nitrogen. After reconnecting the sample tubing, the heavy key concentration was very close to the target, and well below the high limit. The reactor was started successfully.

Hours later the analyzer started showing high concentration of heavy key, but this only underscored the proof of the condensation mechanism.

Conclusion:

The time to begin distillation control engineering is during process design phase. Implementing some or all of the recommendations in this paper will improve system performance in a way that cannot be done after plant startup.

Design teams should begin new projects by statistically analyzing existing similar units, and following the recommendations of this paper to minimize variation and to maximize on-spec production.

References:

1. C. Herzog, 'Develop Distillation Controls During Process Design', CEP Magazine Article, Feb. 2021
2. For a discussion of Statistical Process Control and Quality manufacturing methods and history, Wikipedia has an excellent article. A primary reference book was written by Walter A. Shewhart and edited by W. Edwards Deming, "Statistical Method from the Viewpoint of Quality Control" (1939).
3. C. Herzog, 'Keep Distillation System on Spec at Maximum Rates', May 2018 slide presentation available at South Texas AIChE website
4. C. Herzog, U.S. Patent 'The Heat and Material Balance Method of Process Control' (2004)

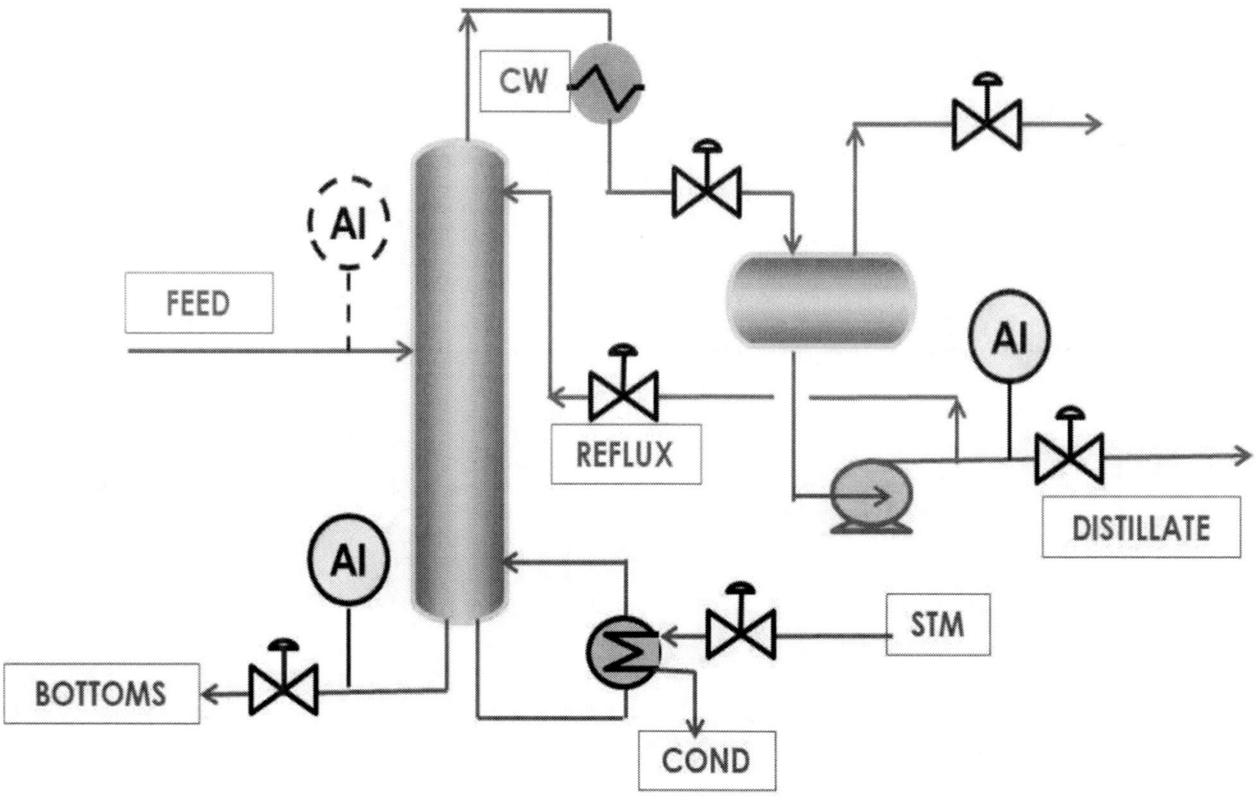

Figure 1

Typical distillation system

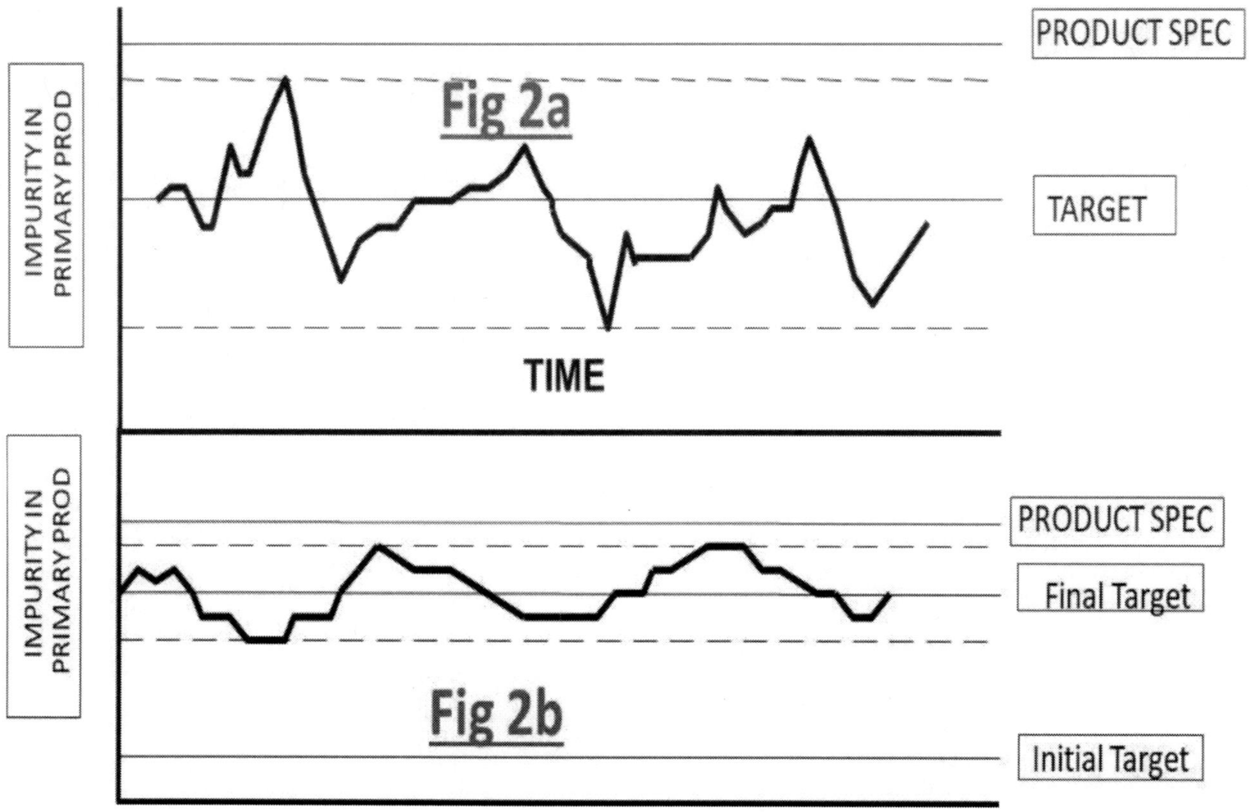

Figure 2: Reduced Variation Leads to Increased Capacity and Better On-Spec Record

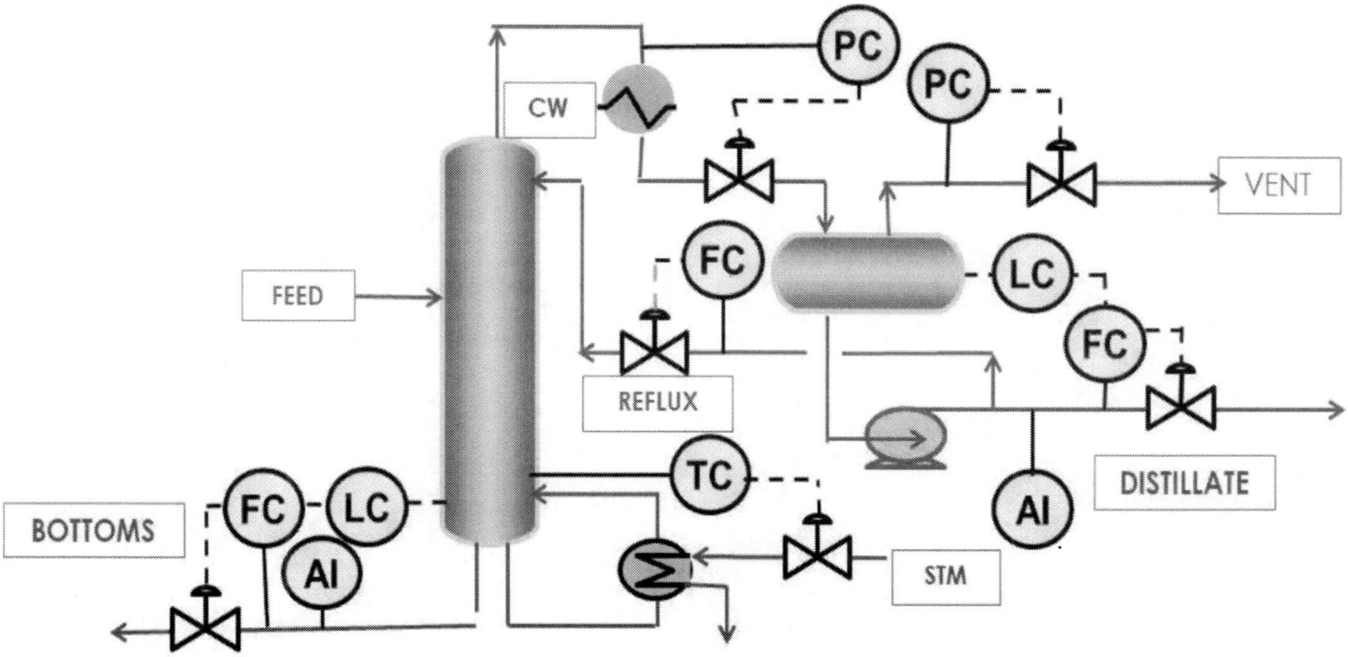

Figure 3: Standard Distillation Control Strategy

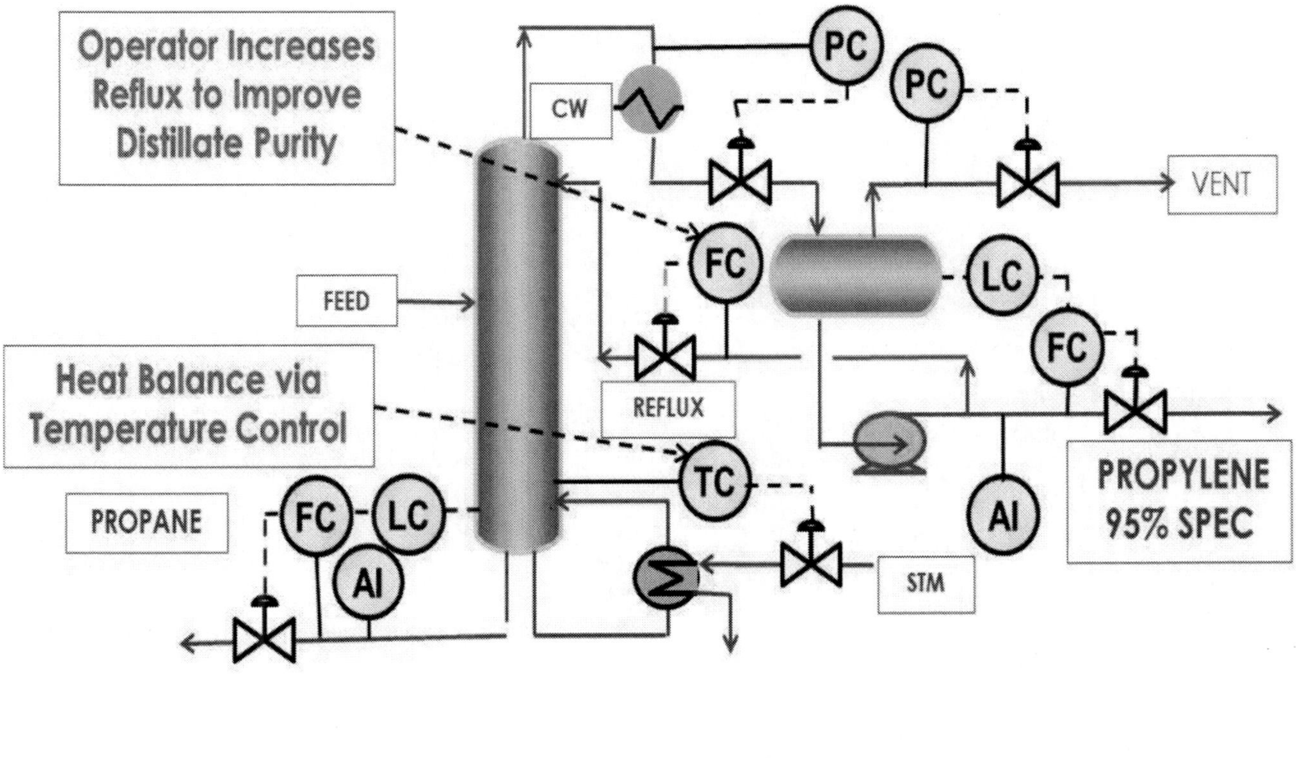

Figure 4: Standard Process Control Strategy as Applied to C3 Splitter

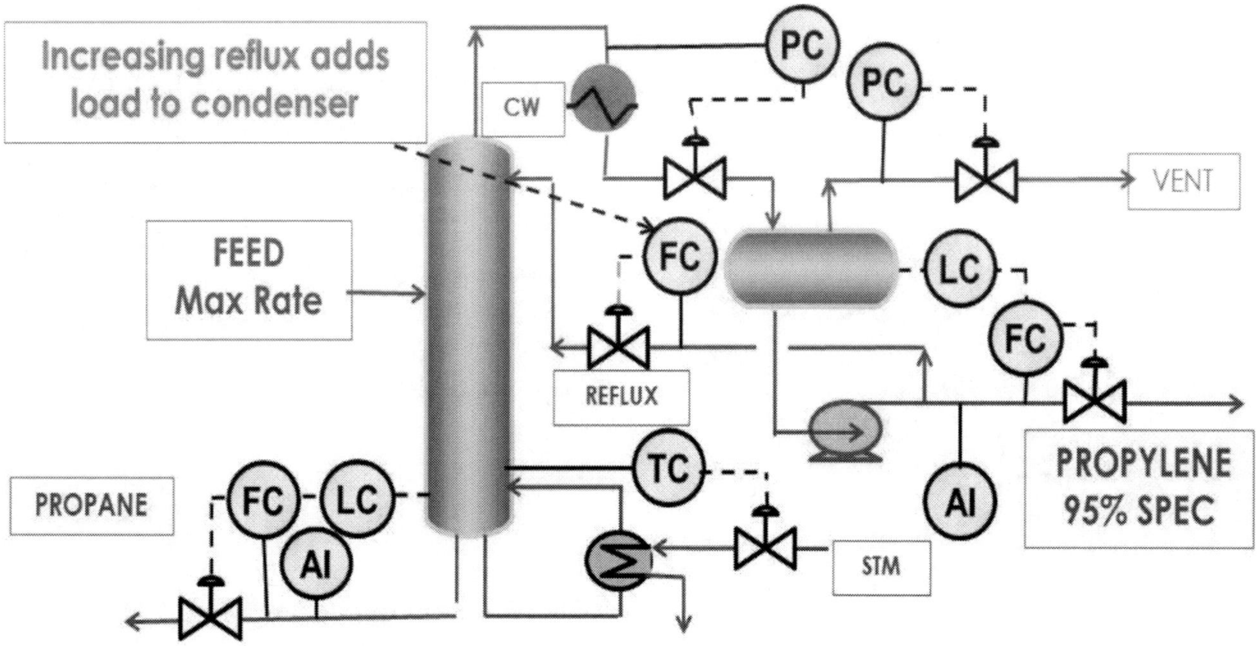

Figure 5: Constrained Condenser Causes Problem for Standard Strategy

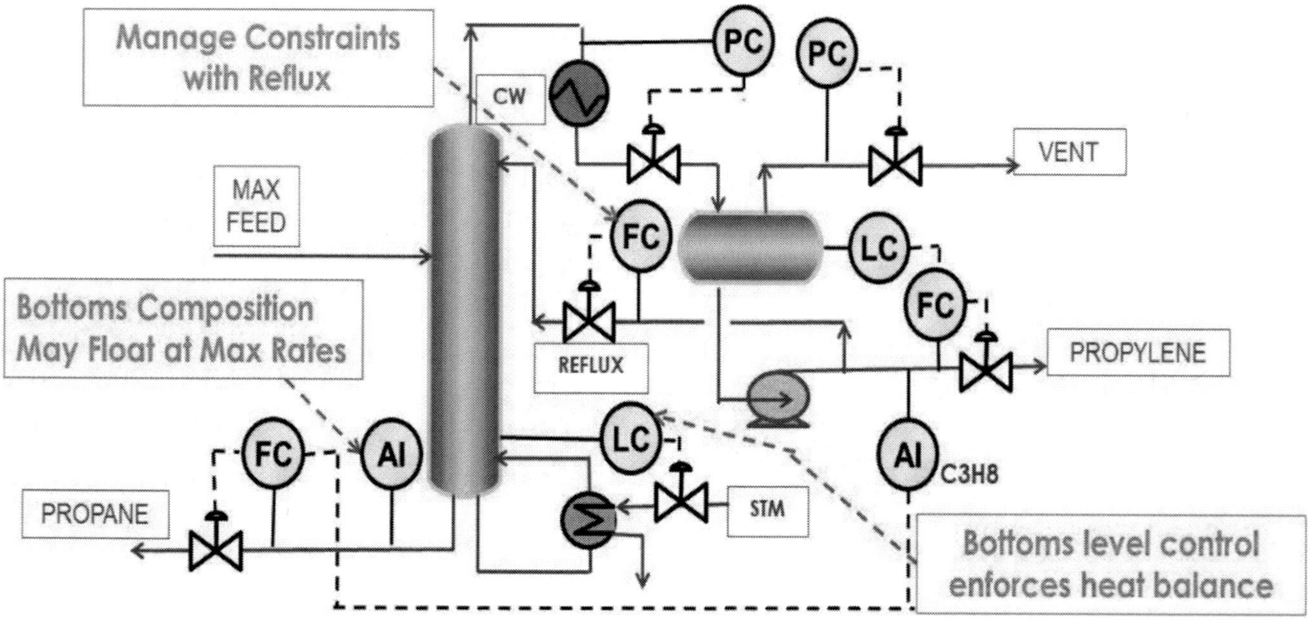

Figure 6: Heat and Material Balance Method Manages Constraints to Keep Product On-Spec

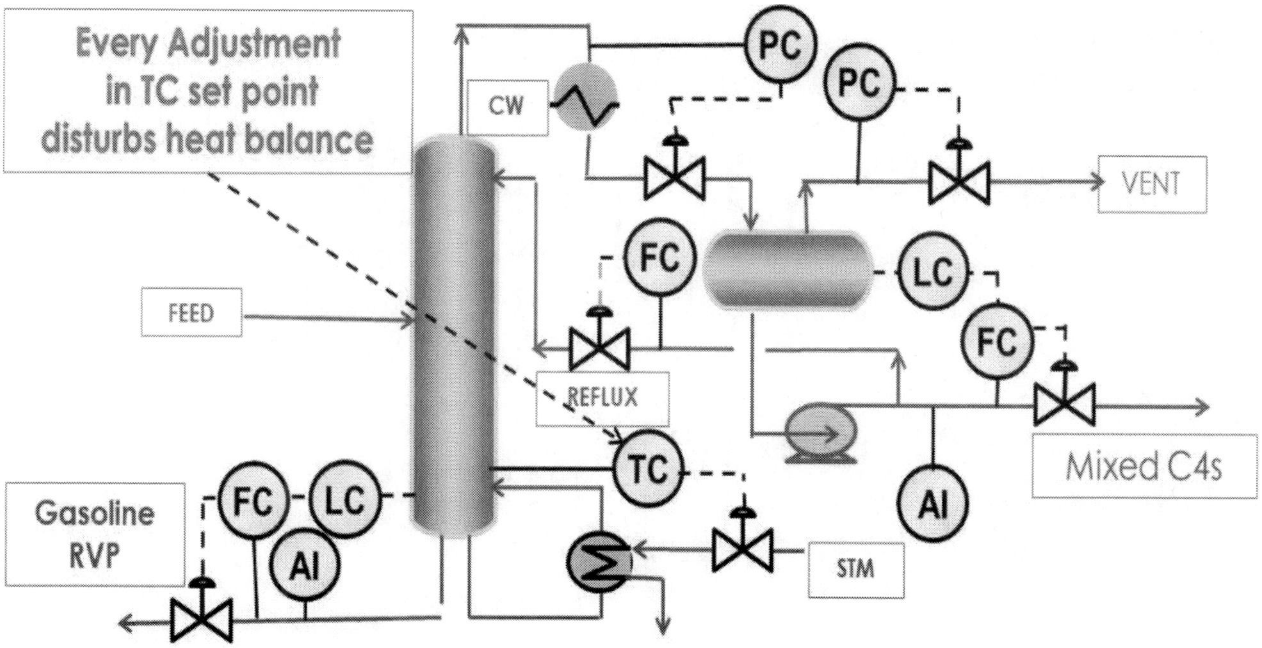

Figure 7: Bottoms Temperature Attempts to Control Gasoline Composition and Heat Balance

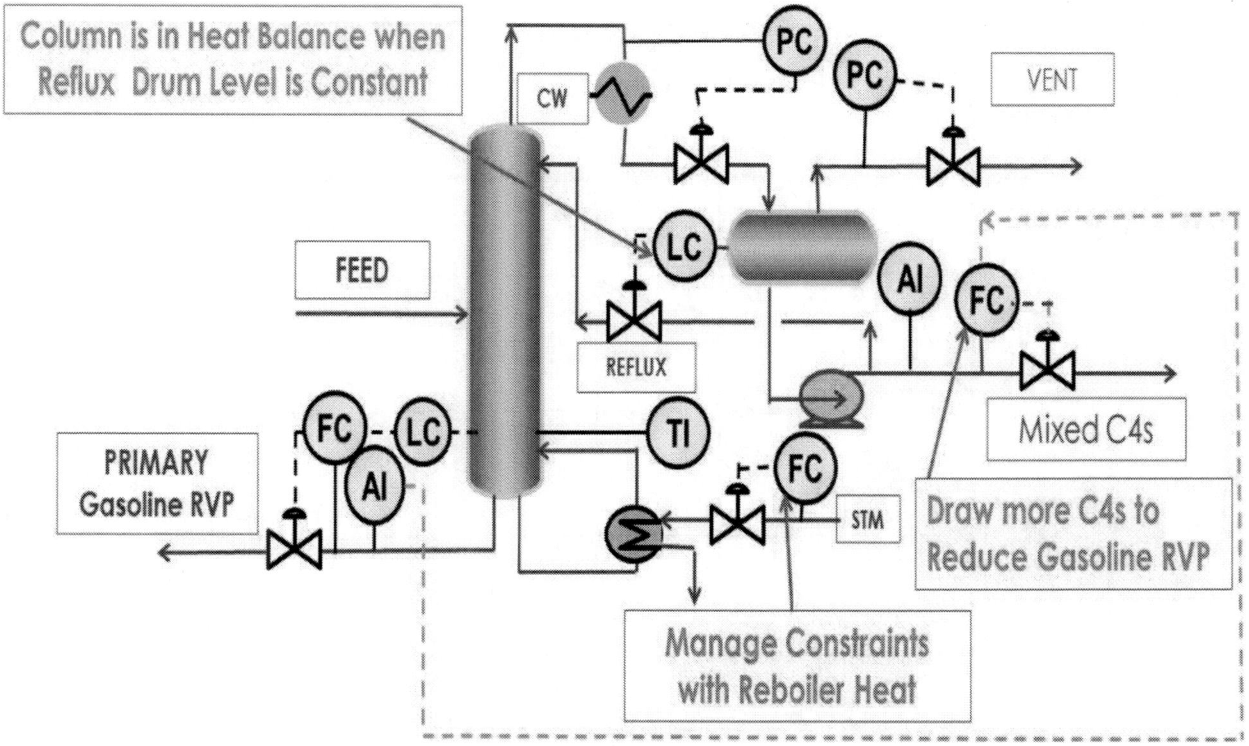

Figure 8: Heat and Material Balance Method Independently Controls RVP and Heat Balance

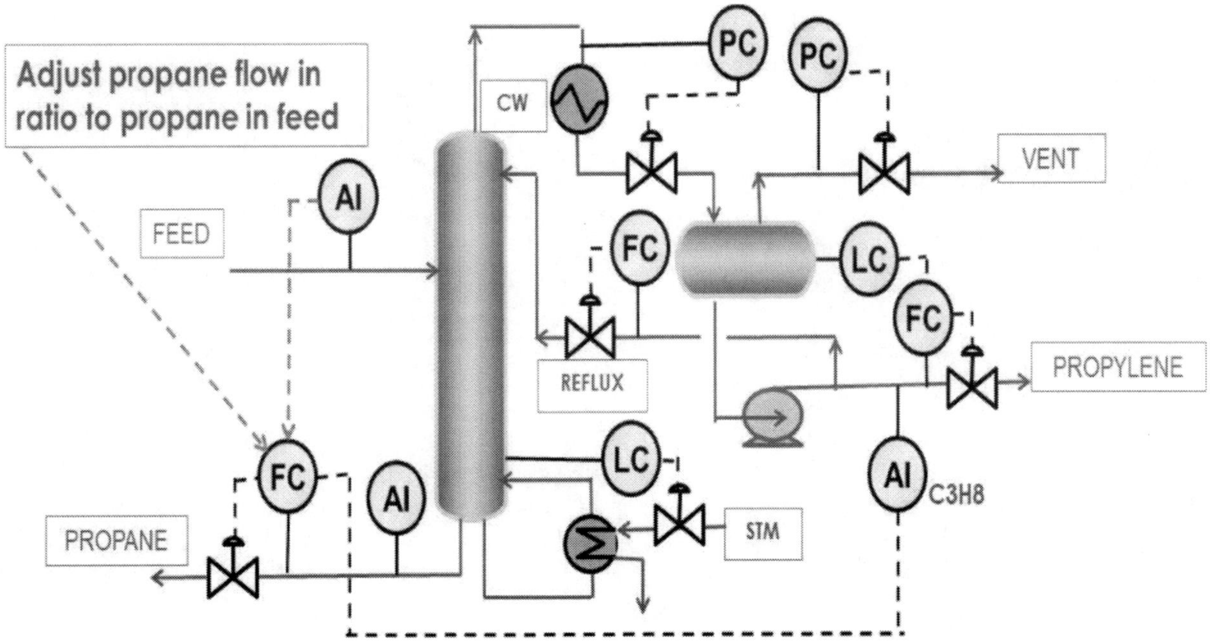

Figure 9: Material Balance Control Strategy Justifies Feed Analyzer

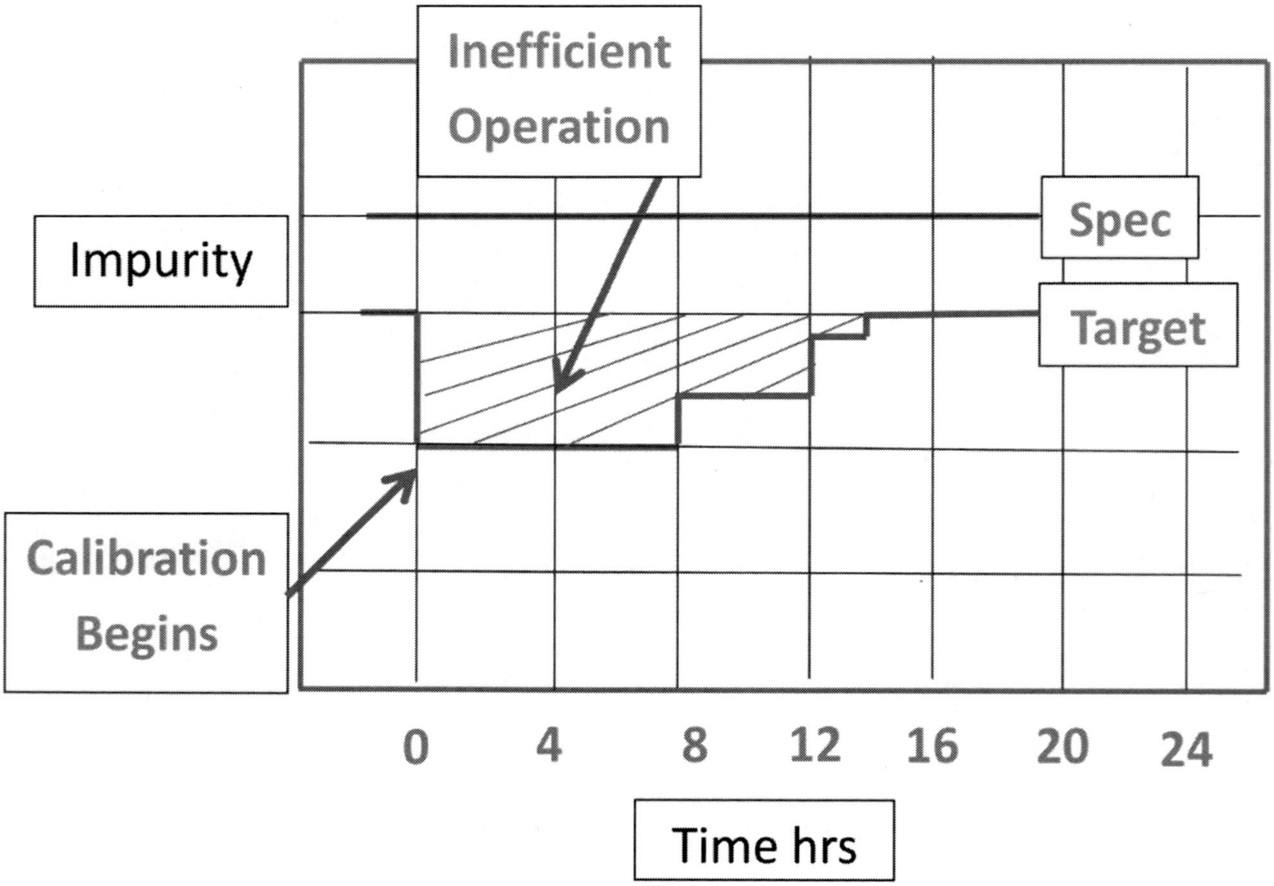

Fig 10: Routine analyzer calibration is a hidden cause of inefficient distillation operations

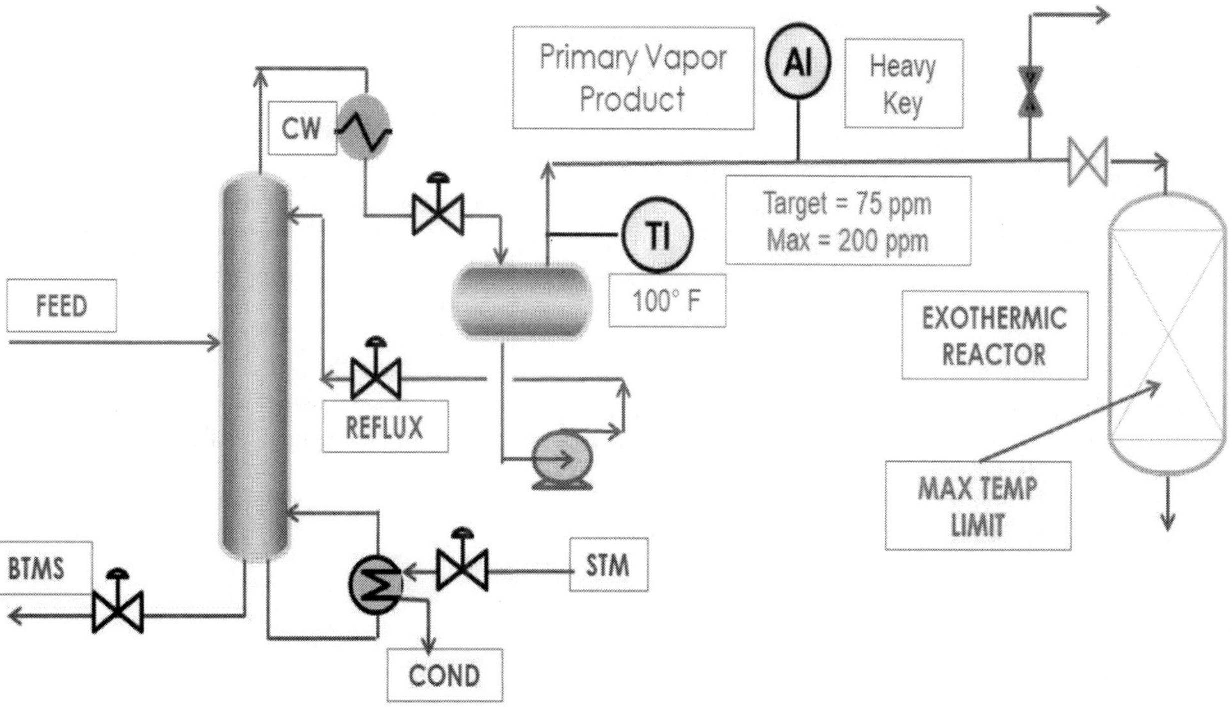

Fig. 11: Critical Analyzer Gives Incorrect Result due to Condensation in Sample Line

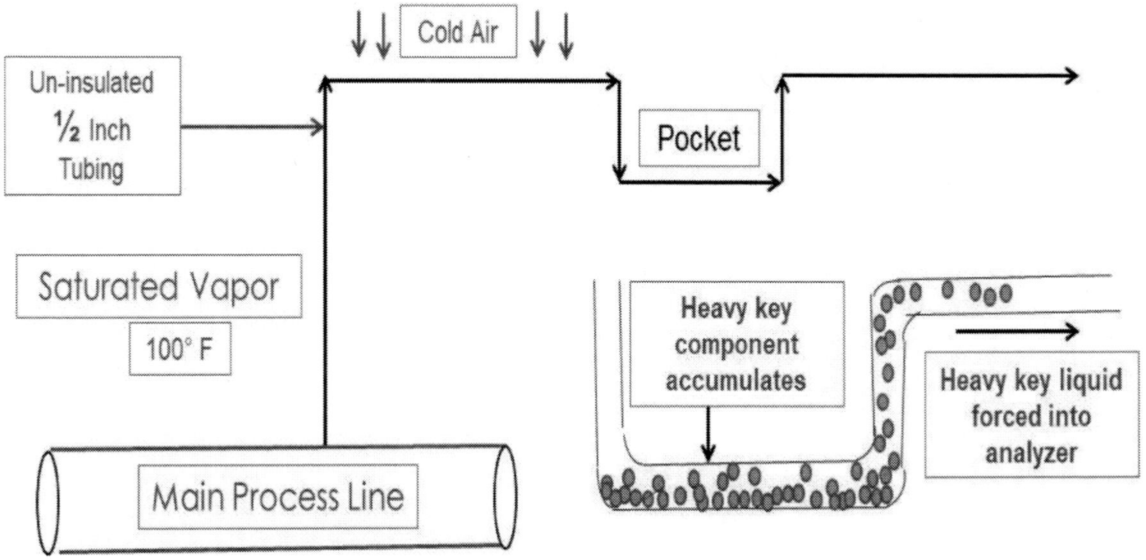

Figure 12: Liquid Rich in Heavy Key Component Forced into Analyzer

62a Meeting New Specifications for CS2 Removal from Petrochemical Naphtha Using Dividing Wall Column Technology

Manish Bhargava, Anju Patil

Economical removal of trace impurities is a big challenge in today's refining and petrochemical complexes. One such challenge in today's scenario is the CS2 contamination which is limiting the feed stock flexibility of petrochemical naphtha (PCN). Presence of CS2 in PCN causes corrosion in downstream units of cracker and also reduces activity of hydrogenation catalysts.

This article provides an insight of how dividing wall column (DWC) technology was implemented in meeting CS2 limit of 2ppm is PCN product at one of the largest refineries in Asia. The DWC was retrofitted in existing depentanizer column. Besides meeting specifications for CS2 removal, the revamped column produced premium gasoline products to boost the revamp profitability.

DWC design required validation of thermodynamic model with plant data and a robust control scheme to maintain high level of product purities.

62b Droplet Entrainment and Separation in Distillation and Absorption Columns

Felix Flegiel, Steven Brown

Both Gas-Carry-Under and Liquid-Carry-Over negatively affect the performance of columns with counter current gas and liquid flow. Especially liquid entrainment in the form of fine droplets often leads to a reduction in stage efficiency, product purity and to hydraulic instabilities such as tray blowing or unsealed downcomers. Insufficient vapor-liquid separation at the column head can result in expensive solvent loss (e.g. methanol or amine) for absorption or wash columns and may even cause damage to downstream equipment, such as adsorbers or compressors. The consequences of insufficient vapor-liquid separation are undesirable both from an economic and ecological point of view. Hence, an optimized column does not only consist of well-designed mass-transfer sections but also requires properly selected and sized phase separation areas and internals.

This presentation covers the basics of vapor-liquid separation in columns such as the Souders-Brown-Equation [1], internals selection, and droplet separation at low surface tensions as encountered in natural gas plants at high pressures [2] or other cryogenic processes. Similarities between capacity limits of wire mesh mist eliminators and tower packings will be discussed. Lastly, some examples from the extensive Linde Engineering plant portfolio will be presented.

References:

[1] M. Souders, G. G. Brown, Ind. Eng. Chem. 1934, 2, 98-103

[2] F. Flegiel, C. Windmeier, M. Wiezorek, A. Döß, M. Schubert, U. Hampel, E. Schleicher, Chemie Ingenieur Technik 2021, 93: 1166-1177

62c Crossflow ULTRA-Frac® Tray Technology

Neil Sandford, Charles Griesel, Emily Ruhl, Anand Vennavelli

Ultra-high-capacity co-current flow trays, such as the ULTRA-FRAC® tray, have excellent capacity but are largely point efficiency devices. The benefits of longer flow path length on the efficiency of conventional and high-capacity crossflow trays is well known. Crossflow ULTRA-FRAC® tray technology blends the advantages of a co-current device's capacity and crossflow tray efficiency to provide a unique solution for towers that need more vapor handling capacity and good efficiency. This paper will discuss recent advancements in the technology, including pilot plant test results and examples of commercial applications.

62e Distillation Tower Startup Following a Plant Revamp with Focus on Reboiler Operation

Andre Bernard

Change in tower internals, twinning of condensers and reboilers implemented as part of an ethylene plant revamp resulted in different behavior requiring adjustments in startup strategies. Pre-startup work established process targets through the feed rate transition with consideration to a number of commissioning hold point requiring period of operation below turndown. On many towers, turndown requirements at low feed rates can be satisfied by increasing internal vapor/ liquid circulation through higher reboil and reflux ratios. Other towers, such as Demethanizer Pre-Strippers cannot be false loaded and can results in reboiler stall. This paper discusses the pre-startup strategy developed and the challenge encountered in the execution of the actual startup. While the discussion covers towers and associated equipment, it focuses on reboiler transient behavior around a Demethanizer Pre-Stripper and two parallel Ethylene Fractionator. The three towers discussed have different sump/ reboiler configurations.

156a Troubleshooting a Methanol Recovery Unit

Christian Geipel

New plants are usually designed for a stable operation meeting all the requirements of the plant operators. However, sometimes it happens that the plant does not operate properly anymore, for example, due to a change in upstream or downstream components and compositions or a change in load range. In this case, a troubleshooting is required to get the column operated again and to ensure product quality at the outlet.

In this context, one of our clients approached us, reporting unstable column operation and unsatisfactory separation of his methanol recovery tower. The tower is divided in two packed beds equipped with high-capacity structured packing made of stainless steel and has so far been operating without any issues. After a regular turnaround of the column, the column does not come back to a satisfactorily operating point. Therefore, a troubleshooting was carried out in close collaboration with the operational team and engineers of the client.

In this study, the different steps of our troubleshooting will be presented. Firstly, a comprehensive simulation study and sensitivity analysis has been carried out to be able to get a first understanding of the issue. Based on the simulation model, former and current column operation modes and set-ups are compared to detect the possible reason for male function. The simulation results show that the HETP of the bottom bed was much higher than initially expected during column design. In addition, when the column was opened up, the findings confirmed the results of our simulation study.

After identifying the root cause, a revamp is developed to get the column working properly again. A two-step revamp procedure is used to overcome bottlenecking of the column and even increase column capacity due to implementing elements of our high-capacity product line.

The study gives a comprehensive insight in the way forward of trouble shooting of distillation columns. The presentation will show the advantage of combination of process simulation calibrated with plant data, operational experience and customized engineering of high-capacity column internals leading to solutions to overcome process issues and increase column capacity.

156b Debottlenecking of a Semi-Batch Vacuum Distillation Column in a Chemical Plant

Claudio Ribeiro Jr., Ann Aisling

Purification of a dilute organic mixture (1 wt% light component) at a chemical plant was carried out by vacuum distillation in semi-batch mode, with continuous feed and bottoms flow but accumulation of the light component in the overhead receiver up to at least 35 times its feed concentration. The column is equipped with a forced-circulation reboiler, a spray condenser, and two beds of 3rd-generation random packing (IMTP). This purification step was the bottleneck for production. A dynamic simulation model was constructed to allow the capacity rating of both the column and its reboiler. High pressure drop in the packed bed, combined with low heat-transfer coefficients in the reboiler due to laminar flow (in the single-phase zone) and film boiling, were found to be responsible for the current capacity limitation. To overcome these limitations, installation of 4th-generation random packing, together with the use of turbulators inside the reboiler tubes and an increase in the recirculation flow rate, were recommended. With the proposed changes, film boiling was no longer flagged in the calculations, and the maximum feed rate to the column increased by 12%.

156e Troubleshooting Cryogenic Distillation Columns

Steven Brown, Felix Flegiel

Shortfalls in observed column separation efficiency and the resulting troubleshootings are a topic that the distillation community is often confronted with. For cryogenic separations, however, there is an added layer of difficulty due to the insulation filled boxes (Coldbox) surrounding the columns, as the options for corrective measures are either limited or come at a higher cost. Further, even small deviations from the design specification can have notable impact on column performance. The necessity to push boundaries of technology, save transportation costs and reduce site work in high-cost countries leads to more shop packed columns and challenging shipping to site. Shop installation can improve quality, but in some cases, there may be a few learnings that are both difficult to discover and even more so to rectify.

This presentation outlines some of the lessons for Linde air separation columns including potential fabrication issues, alignment difficulties and long-distance road shipping. The journey from learning there is a shortfall, diagnosing the potential location that needs remediation, methods to access the columns, the extent of exploration and repair and the overall results will be presented.

Synthesizing a Novel Aromatic Extraction Separation Using Ionic Liquid Solvents

Adam McNeeley[1], Chang-Che Tsai[2], Shiang-Tai Lin[2] and Y. A. Liu[1,]*

**Corresponding author: design@vt.edu; (540) 231-7800.*

[1] AspenTech Center of Excellence in Process System Engineering, Department of Chemical Engineering, Virginia Polytechnic Institute and State University, Blacksburg, Virginia 24061

[2] Department of Chemical Engineering, National Taiwan University, Taipei, Taiwan 10617

1. Introduction

The separation of aromatics from aliphatics in hydrocarbon mixtures using solvents is a practice common in industry. The liquid extraction separation that is best exemplified by the sulfolane process combines integrates three separation steps. A liquid extraction step is used to purify a hydrocarbon of aromatics and obtain an aromatic-rich extract, the aromatic extract is stripped of residual aliphatics in an extractive stripping column, and finally the aromatic and extract are separated in a solvent recovery column and the solvent is stripped of aromatics by a stripping agent. Ionic liquids (ILs) have been investigated as alternative solvents to sulfolane with a key feature being non-volatility. A specific case is made for using IL solvents for low aromatic feeds where the sulfolane process is claimed to be uneconomical. Our goal in this work is to investigate the claims of a non-volatile solvent leading to a fundamentally simpler process and explore possible innovations optimal for an IL process design handling low aromatic content feed.

2. Conventional Aromatic Extraction Process

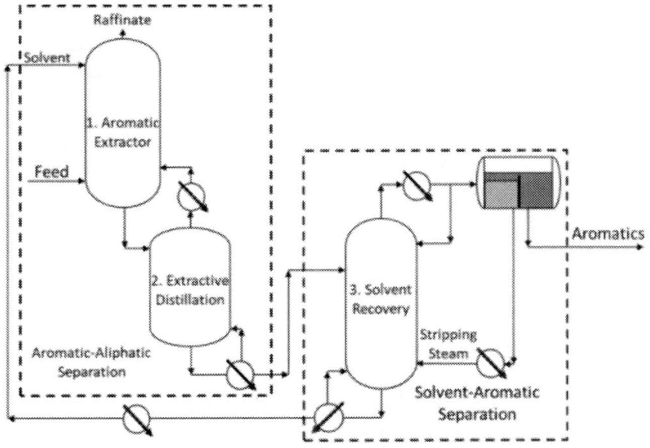

Figure 1. Simplified diagram of aromatic liquid extraction process. Adapted from McNeeley et al.[2]

Table 1. Table of typical purity specification assumed for aromatic extraction process. Adapted from McNeeley et al.[1]

Feed specifications	
Feed rate	50 Mkg/h
Feed temperature	40 °C
Feed pressure	601 kPa
Raffinate product specification	
Aromatic recovery	>98%
Total aromatic content	<0.5 wt%
Solvent content	<100 ppbw
Aromatic product specification	
Benzene aliphatic content	<125 ppmw
Toluene aliphatic content	<1750 ppmw
Mixed xylene alphatic content	<0.5 wt%
Solvent content	<100 ppbw

The initial goal in developing an aromatic extraction process for an IL solvent is to understand the fundamental separation of the conventional liquid extraction process. The most simplified version of the process involves a liquid extraction column where a solvent is feed to the top of the column and a mixed hydrocarbon feed is fed to the bottom of the column shown in Figure 1. The solvent is fed at such a rate that it removes most of the aromatics from the hydrocarbon product at the top of the column. The aromatic extract leaving the bottom of the column contains solvent, aromatics, and residual aliphatics. The aromatic extract is then purified of the residual aliphatics in an extractive distillation column (EDC) by feeding to the top of the column and a purified mixture of aromatics and solvent exists the bottom of the column. The distillate is a mixture of aromatics and aliphatics that is recycled to the liquid extraction column. Finally, the aromatic product is separated from the solvent in the solvent recovery column (SRC).

One common claim of simplification for an IL process is being able to use a single stage flash for the EDC. We consider this simplification to be unrealistic considering the relatively high purity requirements for aromatic products, which we show in Table 1. The required aromatic-aliphatic selectivity to utilize a single stage flash is at least an order of magnitude greater than what the most promising IL solvents can achieve. The design effort in our associated work using a heuristic to select the number of stages for the EDC results in at least 7 equilibrium stages for all the 16 ILs considered[1].

3. Solvent Recovery Column and Its Impact on Process

The solvent recovery section is the area where aromatic product is separated from solvent. A common claim for an IL aromatic extraction process is that the Solvent Recovery Column (SRC) can be a single stage flash due to the non-volatility of ILs. The challenge is the SRC must both produce as solvent free aromatic product, and a relatively aromatic free solvent to recycle. The acceptable amount of aromatic in the recycle solvent is dictated by the liquid extraction and the desired purity of the aliphatic product, raffinate. Figure 2 shows a test varying the concentration of xylene content in the recycle solvent, while controlling a constant aromatic recovery of 98% for a 10wt% aromatic feed composition. We see aromatic content in recycle solvent increases the solvent rate with a limiting concentration at around 250ppmw.

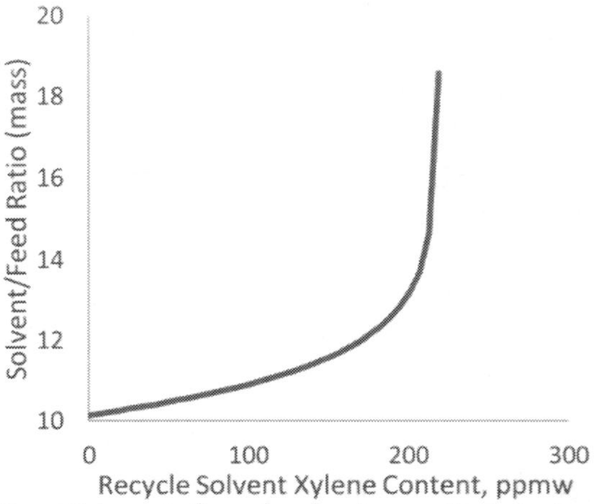

Figure 2. Impact of recycle solvent aromatic content on solvent rate to maintain raffinate purity spec. Adapted from McNeeley et al.[2]

We demonstrate in Figure 3 the solvent composition after the aromatic extract coming from the EDC is subjected to a single stage flash. We reduce the pressure to a minimum value such that the vapor can be condensed at 40ºC and vary the temperature of the flash. Figure 3 shows it is very hard to reduce the composition of xylene recycle solvent with a single flash even at relatively high temperature. We elaborate on this more in our associate work, but in general we find a single stage flash to either require unreasonably high temperatures or relatively low purity raffinate product.

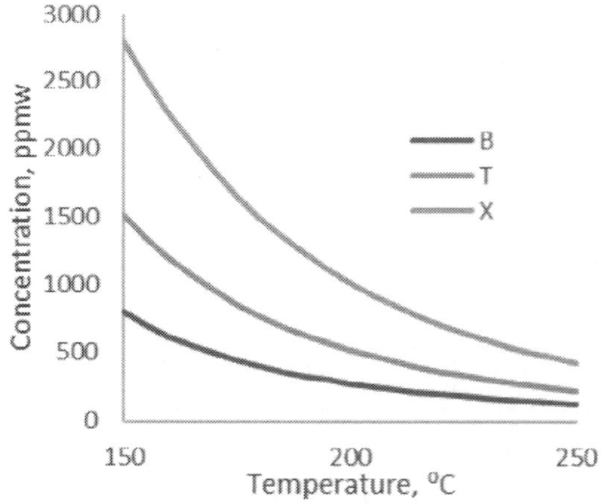

Figure 3. Aromatic compositions (B-benzene, T-toluene, X- p-xylene) in recycle solvent after flash at 0.1bara and varying temperature. Adapted from McNeeley et al.[2]

The solution to be able to control the composition of aromatic in the recycle solvent is to use stripping, which works by introducing a vapor which reduces the partial pressure of the aromatics and thus reduces their composition in the liquid phase. This methodology is already employed by the conventional sulfolane process with steam and we consider a staged steam stripping column to control aromatic content in recycle solvent. The main simplification available to an IL SRC is the removal of the rectifying section and reflux.

We use the limiting concentration of aromatics in recycle as a basis for the control point of the SRC. We use the least soluble aromatic as the basis for the control spec since it is the least volatile and represents the highest contribution of total aromatics in the raffinate product. Simple mass balance can be used to estimate the composition of this component in the raffinate product (x_{Raf}) and the limiting solvent concentration is corresponding equilibrium concentration of this component in the solvent phase (x_{Solv}). The equilibrium concentration can be estimated using activity coefficient ($\gamma_{Raf}, \gamma_{Solv}$) of the component in both the solvent and raffinate phase as shown in Equation 1.

$$x_{Solv} = \frac{\gamma_{Raf}}{\gamma_{Solv}} x_{Raf} \quad (1)$$

$$x_{Set} = ATE * x_{Solv} \quad (2)$$

The control point for the heavy aromatic component in the recycle solvent is then specified using a fractional approach to equilibrium (ATE) as shown in equation 2. This approach ensures the control point for the SRC is less than the limiting concentration of aromatic in recycle solvent and doesn't affect the raffinate purity control. The ATE variable is optimizable for every aromatic extraction process as it balances higher stripping steam rates for lower ATE values with higher solvent rates from higher ATE values.

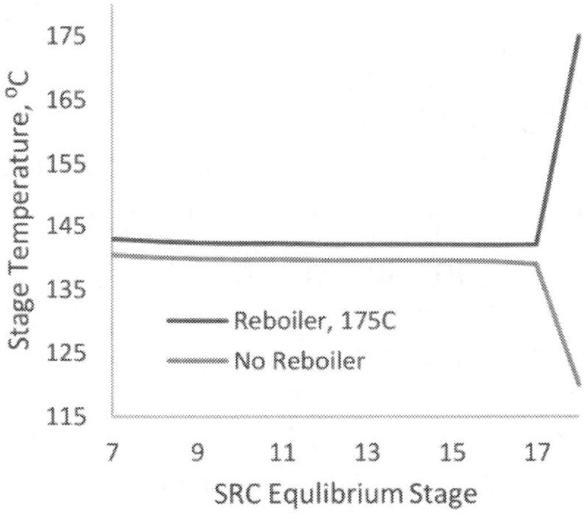

Figure 4. Temperature profile for the SRC for a medium aromatic content feed (60wt%) case with and without an SRC reboiler. Adapted from McNeeley et al.[2]

Literature review of the conventional aromatic extraction process indicates the SRC has a reboiler. We investigated the temperature profile, which is relatively flat as shown in Figure 4. We compared cases with and without a reboiler and found very little difference in the stripping rate or temperature profile in the column. For higher aromatic feed cases an SRC reboiler might be useful for controlling the water content in the recycle solvent, but for low aromatic feed cases the amount or aromatic in the aromatic extract is small and there is very little temperature change from the aromatic adiabatically flashing in the column. Therefore, we eliminate the SRC reboiler for our low aromatic feed case design to simplify the process.

4. Aromatic Extraction Process Improvements

A primary motivation by Meindersma, who originally proposed an IL based aromatic extraction process, for processing low aromatic content feeds is the inability of the conventional sulfolane process to handle low aromatic content feeds[3]. We investigated this claim using a model that is representative of the sulfolane process at varying aromatic feed content using methodologies described in our associate work[2]. We find an optimization variable for this process is the operating pressure of the EDC and that as aromatic content is reduced, the optimal pressure of the column reduces. Eventually the EDC reaches a minimum

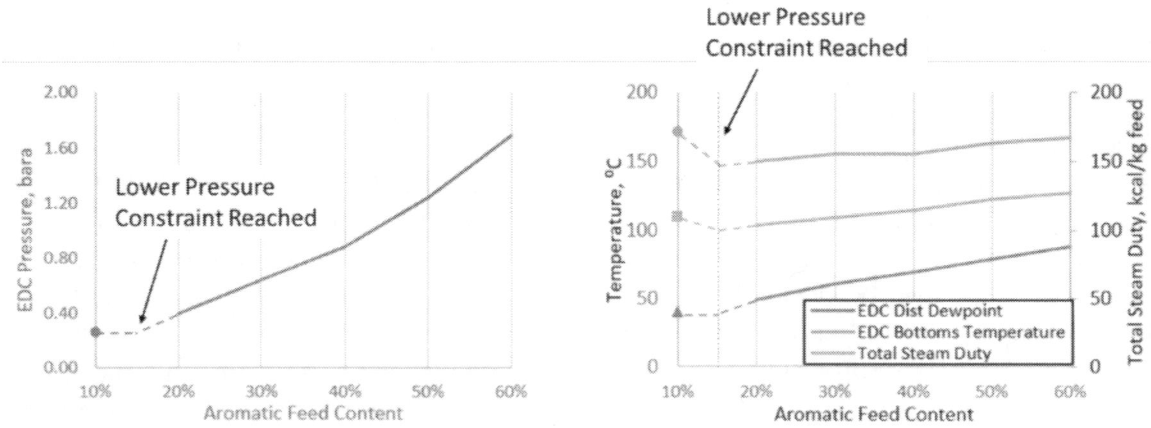

Figure 5. Variations in optimal EDC operating pressure and other process variables for a representative sulfolane process at varying aromatic feed composition. Data obtained from McNeeley et al.[2]

pressure where the EDC distillate is condensable by cooling water. When this minimum pressure is reached, we find the bottoms temperature of the EDC increases drastically and the total process steam duty also increases.

In this preliminary study we neglected trace light components such as pentanes and butanes in the feed. When we added these components in a more rigorous design the minimum pressure increased from ~0.3 bara to ~0.75 bara. The operating temperature of the EDC reboiler at this pressure is ~205ºC, well above the temperature sulfolane is known to degrade at. This leads us to conclude the primary challenge with the aromatic extraction process at low composition is the inability to control the temperature of the EDC reboiler.

We developed a methodology to independently control the EDC reboiler temperature by adding a degree of freedom and controlling the aromatic purity by the vapor rate in the EDC. The control methodology may vary depending on operating conditions and the solvent being used but in the case of ILs where more vapor is being generated by the EDC reboiler than is required to maintain aromatic purity spec, a portion of the vapor can be directed to the SRC condenser. Cases where the vapor rate from the EDC reboiler is less than required to maintain aromatic purity spec, steam can be used from the stripping steam vaporizer to supplement the vapor rate. The two lines used to control the EDC vapor rate are circled in red in Figure 6.

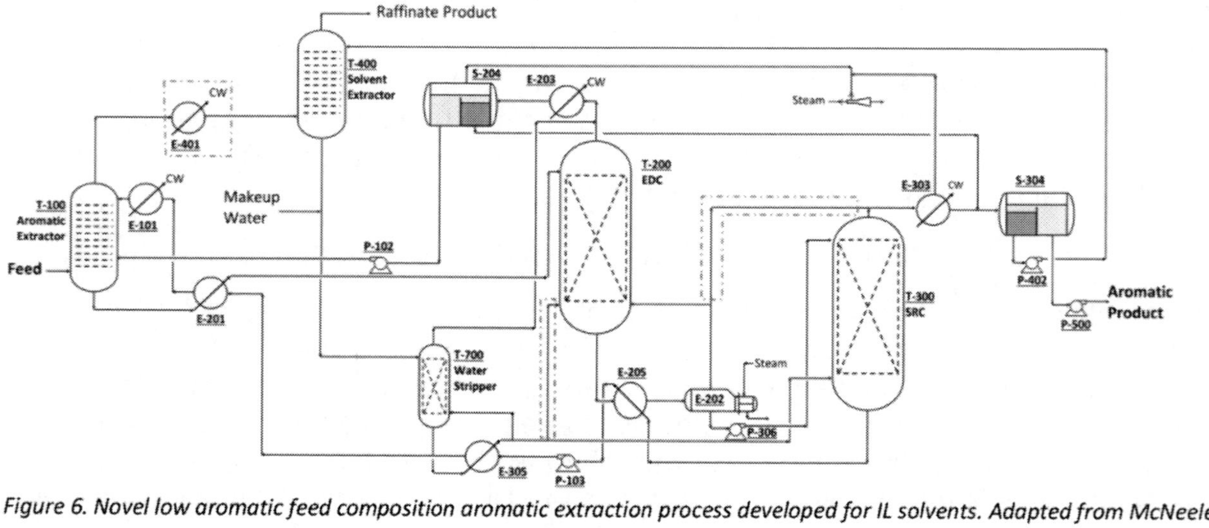

Figure 6. Novel low aromatic feed composition aromatic extraction process developed for IL solvents. Adapted from McNeeley et al.[1]

We add an additional heat integration exchanger that is advantageous for the low aromatic feed compositions where the higher solvent rates lead to a higher solvent heat load to be recovered by heat integration. Conventionally, most of the solvent heat load is recovered by the EDC preheater (E-201). This heat integration is relatively efficient for medium aromatic feed compositions when the preheat temperature is near the bubble point of the feed entering the EDC. When the preheat temperature gets too high the efficiency of the heat integration drops as aromatics flash in the EDC feed lowering its temperature and must be recycled back through the liquid extractor increasing solvent rates. At low aromatic feed compositions, the EDC pressure drops to a minimum value and the bubble point temperature of the feed becomes much lower than the preheat temperature.

Figure 7 shows the how the heat integration efficiency of the EDC preheater is substantially below 1 and decreases as the preheat temperature increases. We define the heat integration efficiency (η) in Equation 3 as the change in total process steam duty (ΔQ_{Reb}) per changer in EDC preheater duty (ΔQ_{PRHT}). The heat integration exchanger (E-205) we add at the bottom of the EDC before the reboiler has a heat integration efficiency of 1 and reduces the amount of heat needed to be recycled to the EDC preheater and can reduce the overall process steam duty by up to 50%.

$$\eta = -\frac{\Delta Q_{Reb}}{\Delta Q_{PRHT}} \quad (3)$$

The novel aromatic extraction process design in Figure 6 has the advantage of having only one major heat input, the EDC reboiler. The EDC reboiler operating temperature is an optimization variable that has substantial influence on the process heat integration. We conducted a sensitivity study on this variable to see how it influenced both process steam duty and heat integration duty in Figure 8. The temperatures that minimized process steam duty are relatively high, but we also find reducing the EDC reboiler temperature doesn't significantly increase process steam duty, while having significant impact on heat integration duty. We decided for our work to use

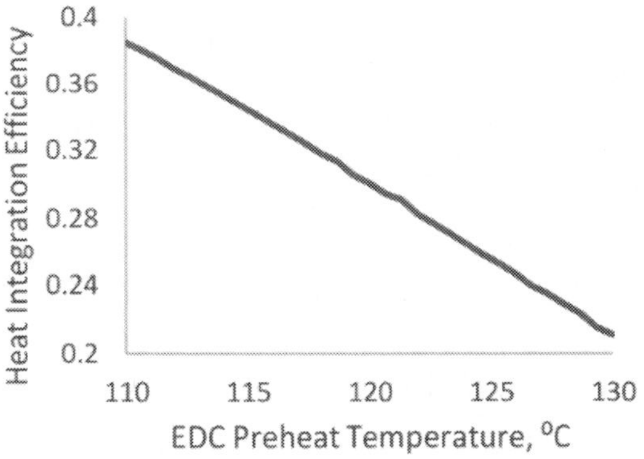

Figure 7. Change in EDC preheater heat integration efficiency at increasing preheat temperature.

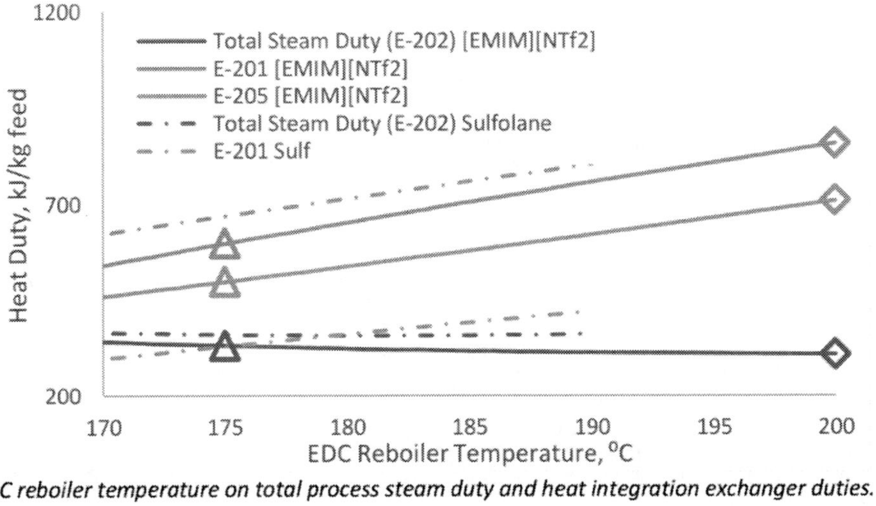

Figure 8. Influence of EDC reboiler temperature on total process steam duty and heat integration exchanger duties. Adapted from McNeeley et al.[2]

at maximum temperature of 175°C assuming the existence of available 150psig steam utility at cheaper cost than higher grade steam and the benefit of smaller heat integration exchangers offset the slightly higher process steam duty.

5. Preliminary Investigation of Extractive Distillation Process Control

A preliminary investigation is conducted to determine a possible aromatic purity control strategy for the EDC. We use the control methodology proposed by Anderson for extractive distillation as the starting point of our analysis where composition of the bottom of the extractive distillation column bottom composition (XB) is controlled using Equation 4[4] where T is a stage temperature in the column, and R/S is the column reflux/solvent ratio and a, b, c are regression constants. Since there is no reflux in the EDC and we find a linear relationship between the log of composition and stage temperature, we initially consider a modified Equation 5 where S/F is solvent/feed ratio.

$$XB = a + b * T + c * R/S \quad (4)$$

$$\ln{(XB)} = a + b * T + c * S/F \quad (5)$$

To pick a temperature control point we compare the relationship between temperature at different stages of a 7 stage EDC and composition for a 10wt% aromatic feed case using [BMPY][TCM] solvent taken from our associated work[1]. Here we find that the temperature-composition relationship can invert at low composition, which is problematic for control and indicates a constraint is likely required to keep the controller in a region where the temperature-composition relationship is non-inverting. We pick temperature at equilibrium stage 3 as ideal for compositional control. We regress coefficients shown in figure 9 over a composition range of 500-5,000 ppmw aliphatics/aromatics and find a good fit as indicated by a good R2 score.

A challenge is that this new process allows independent control of the EDC reboiler temperature and its temperature may be changed and optimized during operation. We find this temperature impacts that predictive capability of Equation 3. We find that the regression constants change as a function of the EDC reboiler temperature as shown in Figure 10, and

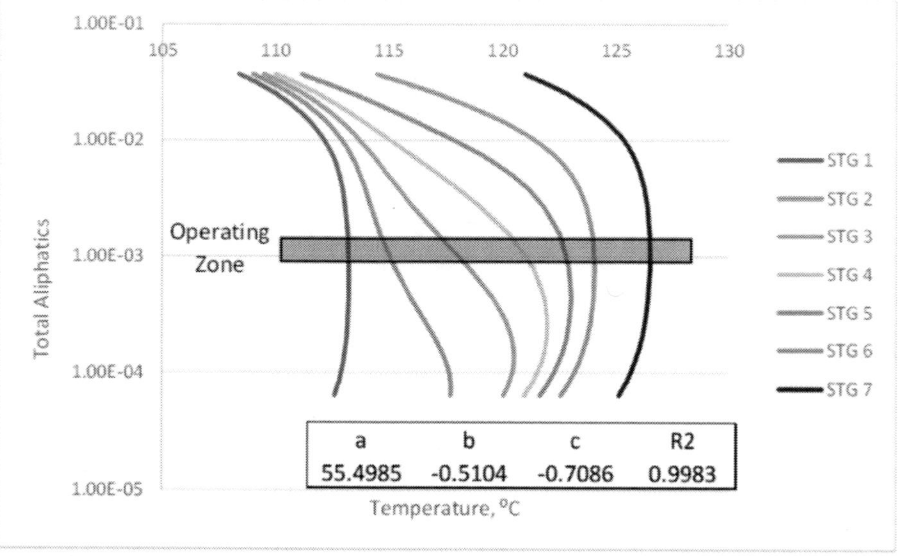

Figure 9. Relationship of EDC temperature profile to EDC aliphatic/aromatic composition. Target control range is ~1,000-2,000 ppmw. Regression parameters and R^2 for control equation, Equation 5, provided using equilibrium stage 3 temperature.

thus we propose a modified form, Equation 6 where the constants (a, b, c) are replaced with functions of EDC reboiler temperature as shown in Equation 7.

$$\ln (XB) = f(T_{reb}) + g(T_{reb}) * T + h(T_{reb}) * S/F \qquad (6)$$

$$f(T_{reb}), g(T_{reb}), h(T_{reb}) = A * T_{reb} + B \qquad (7)$$

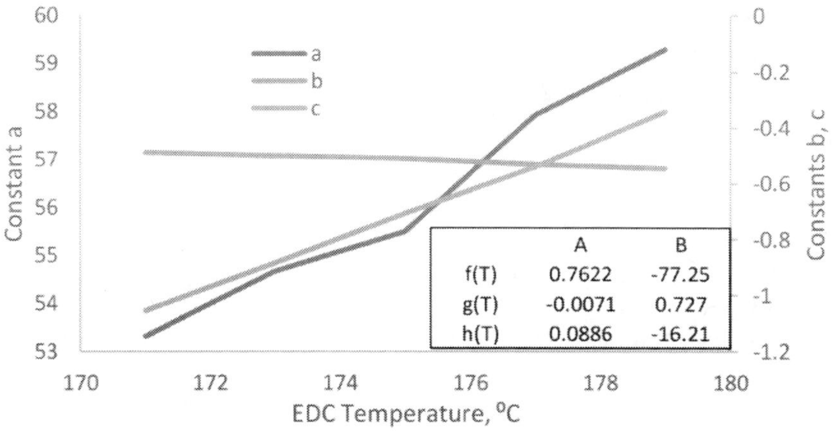

Figure 10. Relationship of regression constants at varying EDC Reboiler temperature. These relationships are used to convert constants into linear functions of reboiler temperature with regression constants (A, B) provided in figure.

We test the effectiveness of Equation 6 to be used to control composition by holding operating conditions constant at base values (EDC pressure 0.49 bara and EDC reboiler 175°C) and perturbing variables such as solvent rate, feed rate, and other process variables shown in Table 2. We find that the equation performs well when varying solvent rate, feed rate, and EDC reboiler temperature, however it seems minor variations in EDC pressure and aromatic feed content have significant impact on composition and may need to be included in a more advanced control model.

Overall, we find that the inversion of EDC temperature at varying composition as well as the sensitivity of the column temperature-composition relationship to minor variations in EDC column pressure and the aromatic feed content to be challenging problems to overcome. We expect these to be well addressed along with other technical challenges as part of the necessary work to transition this conceptual process to commercial operation.

Table 2. Results of steady state process perturbations using Equation 7 as control for aliphatic concentration EDC bottom. All variable except for perturbed variable held constant during simulation. Base case references 10wt% aromatic feed composition, 175°C EDC reboiler temperature, and 0.49 bara EDC pressure.

	Perturbation	Aliphatic Spec ppmw	Control Result ppmw	Deviation
Base		1000	1032	3%
Feed	10%	1000	1023	2%
	-10%	1000	898	-10%
Duty	173°C	1000	1027	3%
	177°C	1000	1035	3%
Solvent	10%	1000	922	-8%
	-10%	1000	997	0%
Pressure	0.51 bara	1000	1697	70%
	0.47 bara	1000	595	-41%
Aromatic	10%	1000	243	-76%
	-10%	1000	3189	219%

6. Conclusion

Our analysis of the fundamental separation steps in the aromatic extraction process and how they are interrelated allow us to develop an energy-optimum process targeted towards the separation of low aromatic content hydrocarbons where the conventional sulfolane aromatic extraction process is not utilized. Overall, we expect an IL based aromatic extraction process to be quite similar to conventional solvent processes with the main difference being the lack of a rectifying section for solvent recovery. The modifications we make in our novel process address issues related to the large solvent heat load and higher extractive stripping column temperatures. We solve these issues by improving the solvent heat integration and introducing a control scheme that allows independent control of extractive stripping column reboiler temperature. We find that the sulfolane process is feasible at low aromatic feed compositions using this new configuration with reasonable process steam duty and our analysis of 16 IL solvents using this configuration reveal up to 46% reduction in process steam duty compared to sulfolane solvent using IL solvents.

7. References

1. McNeeley A., Tsai C.-C., Lin S.-T., Liu Y.A. Science-guided data analytics for selecting ionic solvents for aromatic extraction. *AIChE Journal.* 2023;

2. McNeeley A., Tsai C.-C., Lin S.-T., Liu Y.A. Development of energy-optimum aromatic extraction processes using ionic liquid [emim][ntf2]. *AIChE Journal.* 2022:e17888.

3. Meindersma G.W. *Extraction of aromatics from naphtha with ionic liquids.* University of Twente; 2005.

4. Anderson J.E. Control by tray temperature of extractive distillation. In: Luyben WL, ed. *Practical distillation control.* Springer US; 1992:405-412.

AUTHOR INDEX

AIChE
120 Wall Street, FL 23
New York, NY 10005-4020

ISBN 978-1-7138-7937-4

CURRAN ASSOCIATES INC.
proceedings.com